Robin

Animal

Series editor: Jonathan Burt

Already published

Albatross Graham Barwell · *Ant* Charlotte Sleigh · *Ape* John Sorenson · *Badger* Daniel Heath Justice
Bat Tessa Laird · *Bear* Robert E. Bieder · *Beaver* Rachel Poliquin · *Bedbug* Klaus Reinhardt
Bee Claire Preston · *Beetle* Adam Dodd · *Bison* Desmond Morris · *Camel* Robert Irwin
Cat Katharine M. Rogers · *Chicken* Annie Potts · *Cockroach* Marion Copeland · *Cow* Hannah Velten
Crab Cynthia Chris · *Crocodile* Dan Wylie · *Crow* Boria Sax · *Deer* John Fletcher · *Dog* Susan McHugh
Dolphin Alan Rauch · *Donkey* Jill Bough · *Duck* Victoria de Rijke · *Eagle* Janine Rogers · *Eel* Richard Schweid · *Elephant* Dan Wylie · *Falcon* Helen Macdonald · *Flamingo* Caitlin R. Kight · *Fly* Steven Connor · *Fox* Martin Wallen · *Frog* Charlotte Sleigh · *Giraffe* Edgar Williams · *Goat* Joy Hinson
Goldfish Anna Marie Roos · *Gorilla* Ted Gott and Kathryn Weir · *Guinea Pig* Dorothy Yamamoto
Hare Simon Carnell · *Hedgehog* Hugh Warwick · *Hippopotamus* Edgar Williams · *Horse* Elaine Walker
Human Amanda Rees and Charlotte Sleigh · *Hyena* Mikita Brottman · *Jellyfish* Peter Williams
Kangaroo John Simons · *Kingfisher* Ildiko Szabo · *Leech* Robert G. W. Kirk and Neil Pemberton
Leopard Desmond Morris · *Lion* Deirdre Jackson · *Lizard* Boria Sax · *Llama* Helen Cowie
Lobster Richard J. Kin · *Mole* Steve Gronert Ellerhoff · *Monkey* Desmond Morris · *Moose* Kevin Jackson
Mosquito Richard Jones · *Moth* Matthew Gandy · *Mouse* Georgie Carroll · *Nightingale* Bethan Roberts
Octopus Richard Schweid · *Ostrich* Edgar Williams · *Otter* Daniel Allen · *Owl* Desmond Morris
Oyster Rebecca Stott · *Parrot* Paul Carter · *Peacock* Christine E. Jackson · *Pelican* Barbara Allen
Penguin Stephen Martin · *Pig* Brett Mizelle · *Pigeon* Barbara Allen · *Polar Bear* Margery Fee
Rabbit Victoria Dickinson · *Raccoon* Daniel Heath Justice · *Rat* Jonathan Burt
Rhinoceros Kelly Enright · *Robin* Helen F. Wilson · *Salmon* Peter Coates · *Sardine* Trevor Day
Scorpion Louise M. Pryke · *Seal* Victoria Dickenson · *Shark* Dean Crawford · *Sheep* Philip Armstrong
Skunk Alyce Miller · *Snail* Peter Williams · *Snake* Drake Stutesman · *Sparrow* Kim Todd
Spider Katarzyna and Sergiusz Michalski · *Squid* Martin Wallen · *Swallow* Angela Turner
Swan Peter Young · *Tiger* Susie Green · *Tortoise* Peter Young · *Trout* James Owen
Turtle Louise M. Pryke · *Vulture* Thom van Dooren · *Walrus* John Miller and Louise Miller
Wasp Richard Jones · *Whale* Joe Roman · *Wild Boar* Dorothy Yamamoto · *Wolf* Garry Marvin
Woodpecker Gerard Gorman · *Zebra* Christopher Plumb and Samuel Shaw

Robin

Helen F. Wilson

REAKTION BOOKS

For Esther

Published by
REAKTION BOOKS LTD
Unit 32, Waterside
44–48 Wharf Road
London N1 7UX, UK
www.reaktionbooks.co.uk

First published 2022

Printed and bound in India by Replika Press Pvt. Ltd

A catalogue record for this book is available from the British Library

ISBN 978 1 78914 626 4

Contents

1 A Familiar Bird

'What kind of a bird is he?' Mary asked.
'Doesn't tha' know? He's a robin redbreast,
an' they're th' friendliest, curiousest birds alive.'
Frances Hodgson Burnett, *The Secret Garden* (1911)

For such a small bird, the European robin (*Erithacus rubecula*) has a remarkable portfolio. With its distinctive 'red' breast, it is extraordinarily easy to identify and has a reputation for curiosity, friendship and intimacy that has earned it a sizeable presence in folklore and a diverse set of social and cultural histories. It is said that common species can suffer in popularity on account of being so familiar as to lack novelty; in this matter, the robin has not only avoided such a fate but has emerged triumphant.

The robin is celebrated as a harbinger of seasonal change and a national treasure; it has secured a firm place in a well-established range of Christmas iconography and played a key role in understandings of bird migration and magnetic sense. It has taught lessons in morality and Christian faith, played a prominent role in the founding story of a major city and become a mascot for a variety of environmental campaigns. The popularity of the European robin has seen its name imposed on dozens of birds around the world – scrub robins, bush robins, magpie-robins, birds of all colours and sizes. Some of these robins bear a striking resemblance to the European bird, others less so. Whether the American robin, the fluorescent-coloured robins of Australia or the endangered magpie-robin of the Seychelles, these birds have criss-crossing cultural histories, even if scientifically they are not always closely related. (Throughout the book, and to avoid confusion, 'robin' will

European robin (*Erithacus rubecula*).

James Bolton, *European Robin (Erithacus rubecula) and Eggs, with Wild Strawberry*, c. 1768, watercolour and gouache over graphite, on parchment.

refer to the European robin, and all other robins will be individually introduced.)

Alongside the robin's familiarity, it is the bird's melodious tune and vibrant red breast that have secured it such a prominent social position. The robin is a songbird, as colourful as it is tuneful. Its story is closely tied to the colour red and the colour's rich symbolic histories, which have earned the robin a range of ambivalent associations. Indeed, while known for its companionability, the robin also has a reputation for being a pugnacious bird, a pious soul and a bad omen – a bird of many characters! Robins are known to be both compassionate and violent, solitary and sociable, life-affirming and yet intimately associated with death. While robins have a fairly extensive range, which means they can be found across Europe, North Africa, parts of the Middle East and western parts of Asia, they have an especially prominent history in British folklore and tradition, where the robin has twice been voted the nation's favourite bird. In fact, their association with Britain is so strong that when a vagrant robin turned up in Beijing Zoo in 2019 – only the third time that a robin had been seen in Beijing – it not only drew extensive crowds but prompted a media frenzy and a string of Brexit references.[1] Given the distances involved, it was highly unlikely to have been a British bird, but its quick association with Britain gives some indication of the cultural symbolism that it carries.

It is not only in cultural history that robins have made an impression. It was the robin's familiarity and ease with which it could be caught and ringed that saw it become the focus of a study that was later credited with changing ornithology. David Lack's *The Life of the Robin* was based on field observations at Dartington Hall in Devon, UK, where he was a biology teacher during the 1930s. Lack colour-ringed adult birds and made detailed notes on the function of robin territory, pair formation, migration and the

Christmas postage stamp from Luxembourg, featuring the robin in snow, 2007.

American robin (*Turdus migratorius*).

A robin is examined during a bird-ringing session.

purpose of robin song, to develop a comprehensive account of robin behaviour and ecology across an annual cycle. This was coupled with his observations of captive breeding pairs that were housed in two purpose-built aviaries. Lack's study, which he dedicated to all those robins that 'permitted' his 'intrusions',[2] drew comparisons with the work of J. P. Burkitt, who had undertaken the first study of ringed robins in Northern Ireland ten years before. Burkitt had been the first to use coloured rings to observe birds and record their age – a notable first for the robin.[3]

What made Lack's study especially unusual was its combination of science and culture. Lack's book was praised for both its ground-breaking findings and its ability to speak to scientific and non-scientific audiences alike. As one American reviewer noted in the *Wilson Bulletin*, Lack's book on one of the best-liked birds in 'that bird-minded country' was as much a delightful and absorbing read as it was a 'dependable scientific treatise' and could be enjoyed by both laymen and ornithologists alike.[4] As one of Lack's former students wrote, he taught them that 'the robin was more than a bird; it was a Marco-Polo-type explorer, and a John Keats of song.'[5] It was precisely this movement between science and culture that made Lack's studies so important to the story of the robin and the robin so important to ornithology. Many more of Lack's collected anecdotes and literary references were published as an anthology, which was later extended by his son, Andrew Lack, with a particular focus on the robin's influence on literature.[6] The success of these volumes paved the way for further biographies and continues to demonstrate something of the regard in which the robin is held.[7]

The robin might have a sizeable reputation, but in reality it is a small to medium-sized bird – approximately 14 centimetres (5½ in.) – with a slender bill, predominantly brown plumage and a striking reddish-orange breast. Unlike many other birds, the difference between male and female robins is difficult to discern and while the robin is most famous for its vivid breast feathers, a closer look will reveal that its forehead and face are also adorned with the same reddish-orange feathers, which are bordered with an indistinct band of ash grey.

There are some variations. It is thought that there are eight races of European robin. The variation between these races is somewhat limited but in a line-up it might be possible to detect some differences in the intensity of the face and breast colours, as well as some variation in the colouration of their upper parts. The only exception is the subspecies *Erithacus rubecula superbus*, which can be found on the Spanish Canary Islands of Tenerife and Gran Canaria. This bird has a distinctive pale eye-ring, a much broader band of ash grey, an orange breast that is deeper in colour, and underparts that are almost white. The differences are distinct enough that there has been some suggestion that it might be considered a separate species altogether.[8]

The robin has a striking reddish-orange breast, forehead and face.

A juvenile robin a few hours after leaving its nest.

Unlike the adult bird, you could be excused for misidentifying a juvenile robin. All brown with yellowish streaks and spots, the juvenile doesn't start to develop its adult plumage until it is about ten weeks old, and it will take several more months before it has its full adult plumage. While this process is under way and its reddish-orange feathers are beginning to appear, the young robin can have a rather unusual, slightly scruffy look. Once it has reached maturity, the European robin is unlikely to be confused with anything else, such is the familiarity of its appearance. Indeed, it is not only its vivid breast that makes this bird so recognizable, but its distinctive silhouette. Despite being relatively slim, robins can have an almost spherical shape when their top feather layers have been puffed up to keep them warm, making it look, as it was reported in one newspaper, like they might have swallowed a Christmas bauble.[9]

If shape and colour are not enough, identification is made even easier by the robin's tendency for proximity, a theme which is central to the robin's story and one that has earned it the affectionate title of 'gardener's friend'. It is especially their habit of

Robins are a regular sight at bird tables and will readily take mealworms.

remaining close to any form of activity that might involve the turning of soil that has secured them such a reputation, as I found when taking up paving slabs at the bottom of my garden. As I worked, a robin took up a perch no more than 2 metres (6½ ft) away and remained there for the afternoon, working in sync to scan the ground and snatch up newly exposed worms as the slabs were carried off.

Scanning the ground from a suitable perch is their most common means of foraging but robins will also forage on the ground below bushes, pluck fruit from hedges and take airborne insects. As well as earthworms and slugs, robins feed on woodlice, small grasshoppers and flies, as well as spiders and small molluscs, ants and beetles, while also being partial to some fruits and seeds, including those of apple, bramble, elder, dogwood and mistletoe, to name just a few.[10] In some parts of their range, including Britain, robins are a frequent garden visitor and a regular at the bird table and will feast on fat, sunflower seeds, mealworms and breadcrumbs. However, while the robin might be a familiar garden presence in some parts of Europe, this is not the case across its full range. In Russia it remains a woodland bird and is far less likely to be found in urban areas, while in northwest Africa, robins are largely restricted to montane forests. These exceptions aside, robins can be found across a wide variety of forest, woodland, copse and hedgerow habitats and enjoy the status of being of 'least conservation concern' owing to their healthy numbers.[11]

The robin might be found in diverse habitats, but it was the bird's garden visitations, coupled with its bouncy hops, stops and flits, that won the affections of Frances Hodgson Burnett, who would go on to make the robin the star of one of her most well-known books, *The Secret Garden*. The story follows the sickly Mary Lennox, who is orphaned at a young age and forced to leave her privileged life in India to live with her uncle in a rather grand but

Léo-Paul Robert, *Robin*, 1915, watercolour.

THE SECRET GARDEN
FRANCES HODGSON BURNETT

desolate manor in the middle of the North York Moors. Not long after her arrival she learns of a secret garden that has been locked up and neglected for over ten years. Driven by curiosity she sets out to find it and is helped in no small measure by a robin, who shows her the way.

In the scene where Mary meets the robin for the very first time, the bird is introduced as the gardener's friend. With black dew-drop eyes, curious gazes and 'lively graces', he is frequently seen puffing out his chest, 'as if he were showing her how important and like a human person a robin could be'.[12] In this children's classic, the motif of the garden is linked to the theme of rejuvenation, and as we watch the secret garden come to life, we also follow a year in the life of the robin, who finds a mate, raises a brood and treats the children to the sweetest of songs. He is, as the gardener puts it, the friendliest, 'curiousest' bird alive.

Despite the success of *The Secret Garden*, one of the only records to document the inspiration for the book lies in Burnett's short story *My Robin*, in which the author recounts the circumstance of the real-life robin that became her garden companion in her beloved Maytham Hall, in Kent, where she did much of her writing. *My Robin* was written in response to a letter from a fan. So detailed and joyful were the descriptions of *The Secret Garden*'s robin that the fan could not believe that the bird was 'a mere creature of fantasy' and wished to know whether Burnett owned 'the original of the robin'. In response, Burnett made it very clear that 'one cannot own a robin'. Indeed, if anything, the robin had owned her. For Burnett, the robin was 'a *person* – not a mere bird' and she described him as follows:

> His body is daintily round and plump, his legs are delicately slender. He is a graceful little patrician with an astonishing allurement of bearing. His eye is large and

Front cover of the first illustrated edition of *The Secret Garden* (1911), depicting a robin on a tree branch.

dark and dewy; he wears a tight little satin waistcoat on his full round breast, and every tilt of his head, every flirt of his wing is instinct with dramatic significance. He is fascinatingly conceited – he is determined to engage in social relations at almost any cost, and his raging jealousy of attention paid to less worthy objects than himself drives him at times to efforts to charm and distract which are irresistible.[13]

Burnett's short story detailed shared mornings and companionship, pleasant summers spent among the roses, as well as a final, painful goodbye when she had to return to America and leave her robin behind, never to be seen again. The author's touching description recounted her thrill and wonder for a bird that not only remained in her presence, but seemed to make 'mysterious, almost occult advances towards intimacy'.[14]

Burnett's love for her robin was inseparable from her love for her gardens, which were said to be so spectacular – full of lilies,

The gardener's friend.

The children's character Peter Rabbit, drawn by Beatrix Potter, in the company of robins.

poppies, violets and gigantic roses – that tourists would 'stop in their tracks' to catch a glimpse over the wall.[15] Burnett was writing at a time when gardening had become a national pastime, thanks to new developments in the study and rearing of plants, as was another author: the children's writer and illustrator Beatrix Potter. Potter took great inspiration from her cherished gardens and, like Burnett, also had a keen eye for robins.[16] In her correspondence, Potter wrote of the robin that watched her while she gardened, 'with bright beady eyes, and a very red cap – no, not a cap, a red waistcoat'.[17] While the robin was never given its own book, it did appear in a number of Potter's illustrations, perched on makeshift scarecrows, gathered around Mrs Tiggy-Winkle, and hopping about by Peter Rabbit's shoe in the garden. Perhaps her most famous illustration is of the mischievous Peter Rabbit working his way through a bunch of stolen radishes in the vegetable patch of Mr McGregor. While the young rabbit is undoubtedly the star of the piece, he is accompanied by a robin who perches on the handle of a garden spade, head tilted as though in song.

George W. Temperley (1875–1967) with Robert of Restharrow.

Stories of the gardener's friend abound, but the late ornithologist George W. Temperley was perhaps on friendlier terms than even Burnett or Potter. Upon learning that it was possible to train a robin to feed from his hand, Temperley set about encouraging his garden robin to associate him with food. According to Temperley, the robin, to whom he gave the rather formal name 'Robert of Restharrow', was quick to learn and was soon treating the ornithologist as 'an animated bird table'. Not satisfied with taking mealworms from his cap, Robert was also partial to chewed biscuit from the man's lips. In fact, Temperley remarked that should the biscuit supply run out, Robert would 'flutter impatiently on my finger and chirp until I opened my mouth widely enough for him to plunge in his head to look for the last crumbs which he plucked from my teeth'.[18]

For many, such intimacy might be a step too far, but stories of other robin companions aren't difficult to come by. During the

UK's COVID-19 lockdown, a robin by the name of Colin made the front page of *The Times* after gaining an enviable following on Twitter. His companion, teacher and wildlife enthusiast Kate McRae, documented their burgeoning relationship after Colin first appeared in her garden at the beginning of April 2020. First alighting on her feeding tables, and then gradually inching closer and closer to her, he became a constant companion after their friendship was sealed with mealworms. McRae's documentation of Colin's antics and obliging nature were an instant hit on social media.[19]

Undoubtedly, the robin has long been widely popular, overly familiar and instantly recognizable – and not just within its natural range. As a consequence, it is a bird to which a multitude of associations have been attached, and it's also a bird that has played a sizeable role in a set of quite remarkable histories as a result. But while in some parts of the world it has come to be known as the gardener's friend, beyond this companionable character lies a more complex, contradictory bird that mixes science, culture, legend and faith.

crificium laudis pro
se suisque omnibus. pro
redempcione animarum
suarum. pro spe salutis
et incolumitatis sue
tibi que reddunt vota
sua eterno deo vivo et
vero. Infra canonem.
Communica
ntes et memo
riam vene
rantes: In primis
gloriose semper vir
ginis marie genitri
cis dei et domini nostri
ihesu christi. Sed et beatorum
apostolorum ac martiru
tuorum. Petri Pauli
Andree Jacobi Jo
hannis Thome Ja
cobi Philippi Bar
tholomei Mathei.
Simonis. et Tha
dei Lini Cleti Cle
mentis Sixti Cor
nelii Cipriani Lau
rencii Crisogoni Jo
hannis et Pauli
Cosme et Damia
ni, N. et non et cor
quorum hodie solemp
nis in conspectu glo
rie tue celebratur tri
umphus. Et omni
um sanctorum tuorum
quorum meritis precibus
que concedas. ut in om
nibus protectionis
tue muniamur aux
ilio. Per eundem christum
dominum nostrum. Amen:
Cum magna veneracione.

2 The Global Robin

When is a robin not a robin? This is a question you might ask when stumbling across the claim that the European robin (*Erithacus rubecula*) is the original or 'true' robin. What of all the other robins, false or otherwise? Robins appear in bird lists all over the world. Many of them belong to the same family as the European robin, which includes robin-chats, bush robins, magpie-robins and scrub robins of all colours – black, white, blue, pink and gold. If leaning out of a window in Singapore, you may in all likelihood see the long-tailed Oriental magpie-robin. You might be lucky enough to get a glimpse of the Swynnerton's robin in Zimbabwe, hear the song of the white-tailed robin in the forests of the Himalayas, or catch a sighting of the white-throated robin in central Turkey. Whether red-breasted, blue-fronted, white-headed or olive-flanked, robins are seemingly everywhere.

The European robin is a songbird (oscine passerine), of which there are nearly 5,000 different species, making it the most diverse and successful group of birds in the world. Although sparse, early fossil records suggest that songbirds originated in Australia, with the earliest-known songbird dated to approximately 54 million years ago. But how they came to proliferate is poorly known. Recent hypotheses suggest that diversification occurred within Australia, which was followed by radiations and dispersal out of the Australian region after tectonic collisions opened up a route

Medieval 'ruddock' in the Sherborne Missal, 1399–1407. The bird now known as the robin was first popularly known as the ruddock.

via Asia. This was followed by further diversification and rapid colonization of other continents, including Europe.[1] The timing of dispersal is contested and made difficult by poor fossil records, but the earliest records in Europe are thought to be from the early Oligocene period (approximately 33.9–29 million years ago).[2]

To understand the robin's scientific family, we must first begin with the chats. The chats – scientific name Saxicolinae – to which the robin belongs, are a closely related group of birds that have 'similarities in plumage, behaviour and breeding ecology'. The name 'chat' derives from the sharp alarm calls that are made by some of the most common species within the group and that were among the first to be described or given a colloquial name.[3] Listen to a robin that has spied a cat in its territory and you might hear what sounds like two stones being struck together in quick succession, or marbles being tipped out of a jar. Outside this closely related group of birds, there are several other robins, none of which are related to the chats. This includes the American robin and numerous species of robin in Australia that acquired their names on the basis of perceived similarities with the European robin by settlers and naturalists, including the presence of brightly coloured plumage or a similarity of habit or form. As we shall see, other than playing a part in their cultural history, the robin is not closely related to these birds and is connected in name only.

The family history of the robin is a somewhat complicated affair. While the American robin (*Turdus migratorius*) is now known to be of no relation, this wasn't always the case. In fact, until recently, chats were considered to be a close relative of the thrushes – Turdidae – of which the American robin is a member. Superficially, the commonalities between the chats and the thrushes are relatively clear to see. Members of both groups are known for their beautiful song, tend to be ground-feeders and are often identified by their somewhat bouncy movements. The rise

Oriental magpie-robin (*Copsychus saularis*).

of DNA testing, however, proved these resemblances to be superficial only and the chats, the robin included, were moved to the family Muscicapidae.[4] This family, the scientific name for which derives from the Latin *musca* (fly) and *capere* (to catch), is known as the Old World flycatchers, which is made up of a group of subfamilies, of which the chats are just one.

While families are made up of subfamilies, subfamilies are made up of genera, which are groups of closely related species. In 1758 the robin was one of 554 species of bird to appear in the tenth edition of Linnaeus's *Systema Naturae*, the internationally accepted starting point for the modern scientific naming of animals. The Swedish naturalist, who uniformly applied two-word names to animals and plants, assigned to the robin the name *Motacilla rubecula*.[5] In line with Linnaeus's system, the first word of this 'binomial' name, *Motacilla*, identified the group of species to which the robin was related (its genus) while the second word, *rubecula*, was the name given to the robin to distinguish it from the rest of the group. Derived from the Latin *ruber*, meaning red or ruddy, *rubecula* is still used today, but the robin is no longer assigned to the genus *Motacilla*. Instead, it is assigned to the genus *Erithacus*, which was created in 1800 by the French naturalist Georges Cuvier after many of Linnaeus's original placements were rethought when some of his identified similarities were found to be wrong.[6] Today, the robin finds itself alone, the only species to be placed in *Erithacus* after others in the same group were moved elsewhere.

So the European robin belongs to the genus *Erithacus*, which is part of the chat group of birds, which is placed in the family of Old World flycatchers. But what does this have to do with its claim to being the original robin? This question brings us back to the settlers and naturalists who carried the name around the world, the origins of which lie in medieval English.

The bird that is now known as the robin was first popularly known as the ruddock.[7] The ruddock of medieval English has its roots in the Anglo Saxon *rudduc*, from *rud*, meaning red colour (especially of the complexion). A reader concerned with accuracy might consider ruddock a poor choice for a bird that is more orange than it is red. But what might seem like an oversight can be easy to explain: while the word 'orange' was in use in Middle English to refer to the fruit, it was not used in the English language as a description for a colour that lies somewhere between red and yellow until the early sixteenth century. In any case, as histories of the colour red acknowledge, red has always been a very 'flexible and accommodating colour category' that can cover yellows, oranges, browns and purples.[8] Indeed, the robin's continued association with red might say more about the cultural significance that is attached to the quality of 'redness' than it does anything else.

The name 'ruddock' appeared in Thomas Bewick's *History of British Birds* of 1797, and was still in use in the Victorian period. However, its gradual disappearance from everyday use was already evident in eighteenth-century editions of Shakespeare's *Cymbeline*, where the reference to a ruddock and its 'charitable' bill was accompanied by a short set of explanatory notes for readers who were probably more familiar with the name 'redbreast'.[9] 'Redbreast' – perhaps not the most imaginative of names – appeared from the fifteenth century onwards.[10] It was around this time that affectionate, often alliterative nicknames were also used for some of the most favoured and familiar birds. And so entered Robin Redbreast, sweetheart of Jenny Wren and contemporary of Tom Tit, but while other birds eventually dropped their affectionate names, Robin Redbreast held on to his, eventually becoming simply Robin.

Robin is a diminutive of Robert, a popular name across the European continent during the medieval period. The name was introduced to England by the Normans, but it has ancient Germanic roots. The Old High German *Hrodebert* was a compound of *Hruod* (meaning fame, glory, honour, renown) and *berth*

Thomas Bewick's illustration of a 'Redbreast', 1797. Bewick was a wood-engraver known for his revolutionary technique.

(meaning bright, shining).[11] While the name came to England from northern France, the use of 'Robin' as a bird's name was restricted to the British Isles and was just as likely used because of its popularity than its meaning. There were, however, other affectionate names in circulation. In 1797, the engraver and naturalist Thomas Bewick noted that its general familiarity had occasioned the bird to be distinguished by other peculiar names, including Tomi Liden in Sweden, Thomas Gierdet in Germany and Peter Ronsmad in Norway.[12] Reference to some of these names can be found in William Wordsworth's poem 'The Redbreast and the Butterfly':

Art thou the bird whom man loves best,
The pious bird with the scarlet breast
Our little English Robin;
The bird that comes about our doors,
When Autumn winds are sobbing?
Art thou the Peter of Norway boors?
Their Thomas in Finland.
And Russia far inland?
The bird who by some name or other,
All men who know thee call their brother,
The darling of children and men?[13]

These names are repeated elsewhere in other poems, school lesson books and bird guides, but it is unclear from where these affectionate names originate.[14] In Reverend Smythe Palmer's 1882 dictionary of 'verbal corruptions or words perverted in form or meaning, by false derivation or mistaken analogy' – not a title that rolls off the tongue – he suggests that Norway's Peter Ronsmad might have been a perversion of the bird's name in southern Europe: the Italian *pettirosso*.[15] The literal translation of *pettirosso*,

Anselmus Boëtius de Boodt, *Robin (Erithacus rubecula)*, 1596–1610, watercolour and ink on paper.

just like the Finnish *punarinta*, is 'red breast', while the Norwegian *rødstrupe*, Danish *rødhals*, German *rotkehlchen* and French *rouge-gorge* can be translated as 'red throat'. The Swedish *rödhake* translates more closely as 'red chin', while the Icelandic *glóbrystingur* means 'glowing breast'. While subtly different, what is clear from these names is that the bird's red plumage is its defining feature.

Having established how the robin got its name, it is possible to ascertain how the name travelled. Carl Linnaeus's *Systema Naturae* was published at a time of fervent European exploration. Knowledge was power, and the systematic description, classification and ordering of the world was central to imperial endeavours. It is during this period that the robin appears in the notes and diaries of some of Europe's most notable explorers and naturalists, who were sent on missions by their governments to develop applied knowledge of their colonies and other lands. Scientific exploration was of intense public interest, and travel writing, letters and other forms of documentation played a crucial role in shaping how European readers imagined the world beyond Europe.[16]

For instance, the European robin made an appearance in the journals and notes from Captain Cook's voyages on the *Endeavour*, which were undertaken, by the order of King George III, in order to make discoveries in the southern hemisphere. In 1769, in a section on the religion of the people of Otaheite (now the Polynesian island of Tahiti), it was noted that there had been no evidence of 'idolatry' found on the island, but that the people did have a peculiar regard for a particular bird:

> This island indeed, and the rest that lie near it, have a particular bird, some a heron, and others a king's fisher, to which they pay a peculiar regard, and concerning which they have some superstitious notions with respect to good and bad fortune, as we have of the swallow and robin-red-breast.[17]

A similar reference can be found five years later in an entry from May 1774 on the island of Raiatea, also in Polynesia:

> The people of this isle are in general far more superstitious than at Otaheite or any of the other isles. The first Visit I made the chief after our arrival, he desired I would not suffer any of my people to Shoot the Heron's and Wood Pickers, Birds as Sacred to them as Robin Red-breasts [and] Swallows . . . are to many old women in England.[18]

Of course, the comment that superstitions, and the robin's sacred status, were a matter of concern only among 'old women' couldn't be further from the truth and is indicative of the rejection of sentimentality that characterized the European Enlightenment.

These historical references are not only evidence of the regard in which the robin was held but demonstrate how explorers made sense of the world through their own frames of reference and perceived likenesses to familiar species back at home. These descriptions allowed audiences in Europe to understand something of the attachments that Cook and others had observed elsewhere.

Alongside social attachment, the robin also appears in notes concerning perceived commonalities of character and appearance. For instance, in his account of the common blue-bird (another member of the chat family, and now known as the eastern bluebird), the American naturalist J. J. Audubon noted how the bird reminded him of the robin redbreast of Europe, to which, he suggested, 'it bears a considerable resemblance in form and habits':

> Like the Blue-bird the Red-Breast has large eyes, in which the power of its passions is at times seen to be expressed. Like it also, he alights on the lower branches of a tree, where, standing in the same position, he peeps sidewise at the objects beneath and around . . . Perhaps it may have been on account of having observed some similarity of habits, that the first settlers in Massachusetts named our bird the Blue Robin, a name which it still retains in that State.[19]

The tendency for settlers to name species after those familiar to them is widely acknowledged, as is the strong attachment that was often formed with birds that reminded them of home. As the American ornithologist Oliver Luther Austin Jr put it, 'The Robin has always been so well liked in Britain that Britons have carried the name wherever they went.'[20] In his notes on the scarlet-breasted robin of Australia, the English ornithologist and illustrator John Gould remarked:

Eastern bluebird (*Sialia sialis*), North Carolina.

> When far removed from our native land, recollections and associations are strong incentives to attachment for any object that may remind us of our home; hence this beautiful Robin, which enters the gardens and even the windows of the settlers, is necessarily a great favourite.[21]

While considered feebler in sound, the scarlet robin (*Petroica boodang*) was also noted to have a song and call that much resembled that of the European robin. A medium-sized robin, the male has a scarlet breast with a black throat, while the female has grey upper parts with a lighter-coloured breast. Given the vibrancy of its feathers, it doesn't take much imagination to see why the bird might have attracted favourable comparisons to the familiar bird back in Europe.

The scarlet robin is part of the Petroicidae family of birds, commonly known as the Australasian robins, and is of no close relation to the European robin. While often found in the eucalyptus forests and woodlands of south Australia, it can disperse into open parks, grasslands and gardens in the colder months. The red-capped robin (*Petroica goodenovii*) of the same family was also noted to bear resemblance to the European robin on account of possessing a 'peculiarly sweet and plaintive song'.[22] It is slightly smaller than the scarlet robin but similar in appearance, with a vivid red breast and forecrown, and a widespread inland habitat. Just as striking are the pink robin (*Petroica rodinogaster*) and rose robin (*Petroica rosea*), which were observed to 'present their full breast' much like the bird of Europe. Both are found in wet forest and rainforest, with a tendency to move to drier woodland in the winter. These are just a few of the many birds that inherited their English names from the European bird.[23]

Alongside the tendency to form strong attachments with species that appeared familiar, the desire for the sights and sounds of home was most evident in the activities of acclimatization societies, which were established in the middle of the nineteenth

Scarlet robin (*Petroica boodang*), New South Wales, Australia.

Scarlet robin.

Pink robin (*Petroica rodinogaster*), Tasmania, Australia.

century. These societies were founded with the intention of remaking the landscape of settled lands by introducing (or acclimatizing) birds, animals and plants that in some way recalled their members' home countries. In what has been described as one of the most perverse examples of ecological imperialism, 'home' was recreated in newly settled countries through stocking the fields, woods, plains and forests with European game and songbirds.[24] Of course, what constituted a beautiful and homely landscape was based on a distinctly European idea of nature so there's no surprise that it was the birds that featured most prominently in the European cultural imagination that also featured in acclimatization programmes. The robin, along with the nightingale and the skylark, was high up on the list of birds to ship to the Antipodes, although very few would survive the journey from Europe.

The nostalgia for home was perhaps strongest in Australia and New Zealand. Environmental historian Thomas R. Dunlap notes that, by the mid-nineteenth century, Gilbert White's *Natural History of Selborne* had become a classic and was carried all over the world by homesick Englishmen, who longed for the nature it described.[25] Widely considered to be a nostalgic evocation of a pastoral vision and a precursor to modern ecology, White's natural history offered detailed notes on the behaviour of animals and birds in his rural Hampshire parish in England. It depicted a settled life and social order that was strongly connected to the changing seasons (although, incidentally, White offered a somewhat derisory assessment of 'redbreasts' on account of their tendency to 'do much mischief' to the summer fruits in his garden).[26] It was argued by Frederick McCoy, first director of the National Museum, that English songbirds and their 'varied touching joyous strains of Heaven taught melody' were therefore not only delightful reminders of the English countryside (which

Detail from the Gilbert White memorial in the Church of St Mary, Selborne, which depicts St Francis preaching to all of the birds mentioned in White's writings. The robin perches on the saint's hand, 1920.

was so vivid in White's writing), but that they would address the 'savage silence' of the Australian bush.[27] Such derisory assessments of the rich and diverse landscapes of Australia and New Zealand were rife; the consequences of which continue to be felt today.

When the Acclimatisation Society of Victoria was founded in Melbourne, Australia, in 1861, the society's first president, Edward Wilson, argued that Australia's indigenous animals were practically useless.[28] On the basis of this dire evaluation, the Society's inaugural meeting was full of ambitious promises which were met with applause, not least that tables would very soon be 'groaning with all the delicacies which can be procured in the markets of London and Paris'.[29] In its third annual meeting, it was reported that, on account of a local problem with caterpillars, the introduction of 'insect-destroying birds' had been 'carefully attended to', along with a combined 'effort to surround our colonial residences with those interesting associations which constitute no slight portion of the charms with which the name of "home" is ever surrounded'. Three years after the society's founding, it was reported that such birds as the skylark, the sparrow, the chaffinch and the starling were now 'considered thoroughly established, and rapidly extending by natural means through the colony'. Meanwhile, robins, along with many other smaller birds from other countries, were accumulating and breeding in the society's aviaries, and many more had been promised by 'benevolent ladies at home'.[30] The queen herself had offered to supply a rook!

At the same time, in a bid to encourage settlers of Canterbury to establish their own society, an editorial in Christchurch's newspaper, *The Press,* argued, 'There is perhaps no country in the world the natural zoology of which supplies so little to the subsistence or enjoyment of its inhabitants, as New Zealand.'[31] In 1864 the Canterbury Horticultural and Acclimatisation Society was formed and within a year the robin had appeared on its purchase and introduction lists. Like many other species, the European robin proved impossible to establish.

In a report on the wild birds introduced or 'transplanted' into North America, it was noted that the robin redbreast of Europe had been the subject of 'introduction experiments' on several occasions. Robins were most notably released in the state of Oregon by the Song Bird Club, which had been founded by the German-American C. F. Pfluger with the specific aim of importing European songbirds. Robins were similarly released by the Cincinnati Acclimatization Society in Ohio, which spent $9,000 on experiments involving twenty different species.[32] Robins were also released in Detroit, Michigan, and on a large scale in Victoria, British Columbia, in 1913. Rather peculiarly, in the late nineteenth century the robin was also thought to have been released in Central Park in New York as part of a project to introduce into

America every bird ever mentioned by Shakespeare. This had been the ambition of Eugene Schieffelin, a drug manufacturer and Shakespeare enthusiast, who also happened to be a member of the American Acclimatization Society of New York. The robin was only mentioned twice by Shakespeare (in *The Two Gentleman of Verona* and *Cymbeline*), so it is perhaps fitting that in the history of the robin this New York tribute to Shakespeare is a rather quirky detail of little consequence.[33] In fact, in all instances, the attempts to bring the 'useful song' and beauty of the European robin to the fields and hedgerows of North America were deemed a failure and consigned to the records.

The failed 'transplantation' of the European robin might have been a disappointment for acclimatization societies, but there were other birds that were declared a triumphant success. This included the European starling and the European sparrow, two birds that were initially celebrated for their rapid increase and quick adaptation but quickly vilified when they were perceived to be out of control and a threat to native flora and fauna, including the American robin.

Here is another bird whose history overlaps with the European robin but is not of any close relation. In a diary entry from 12 March 1749, Pehr Kalm, a Finnish botanist who was commissioned by the Royal Swedish Academy of Sciences to travel to North America to collect seeds and plants, noted that the bird which the English and the Swedes in North America called Robin-red-breast 'is a very different bird from that which in England bears the same name'.[34] Pehr Kalm was describing the American robin, a member of the thrush family (Turdidae) and the largest of the North American thrushes. At around 25 centimetres (10 in.) in length, it is nearly twice the size of the European robin.

While not a close relation of the European robin, and in many ways quite distinct, it would be difficult to ignore some similarities.

Described as 'one of the most familiar backyard birds of temperate North America', it is known for its companionability and full-throated sweet song.[35] American robins are common grey-brown birds with a burnt orange underside and can often be seen pulling up worms from lawns, parkland and fields across the continent. While some American robins will spend the winter in their breeding range, many migrate south to warmer states and as far as Central Mexico in search of more abundant food sources. As a result, upon their return for the breeding season, they are often celebrated as a harbinger of spring. The bird's popularity was in evidence in Rachel Carson's influential *Silent Spring*, in which the marine biologist and conservationist documented the destruction wrought by the widespread use of pesticides in 1950s America, especially the agricultural use of the chemical compound DDT.[36] When describing parts of the country where the American robin had been wiped out, or nearly 'annihilated' by the use of chemicals, Carson suggested that the story of the robin might serve as 'the tragic symbol' for the fate of birds:

> To millions of Americans, the season's first robin means that the grip of winter is broken. Its coming is an event reported in newspapers and told eagerly at the breakfast table. And as the number of migrants grows and the first mists of green appear in the woodlands, thousands of people listen for the first dawn chorus of the robins throbbing in the early morning light. But now all is changed, and not even the return of the birds may be taken for granted.[37]

In a survey of newspaper columns and correspondence from readers, Carson pulled out one letter in which a Wisconsin woman asked, 'Can anyone imagine anything as cheerless and dreary as a springtime without a robin's song?'[38] The book brought the issue

John James Audubon, *American Robin*, engraved, printed and coloured by R. Havell, 1832, hand-coloured engraving and aquatint on Whatman wove paper.

of environmental destruction to the public's attention in the face of fierce criticism from chemical manufacturers, resulting in numerous changes to environmental policy.

American robin locking-pin.

The popularity of the American robin was further emphasized by the American ornithologist Oliver L. Austin Jr, although perhaps with some exaggeration. After pointing out that robins had historically been shot for food and sport (you could apparently buy a dozen for 60 cents in 1913), Austin suggested that in contemporary America, 'most Americans would as soon think of eating the family dog' as they would eating a robin.[39] What might be an overstatement aside, this indicates something of the regard in which the American robin is held. In fact, it is so loved and familiar that it is the state bird in Connecticut, Michigan and Wisconsin, and has been described as being as American as apple pie, baseball and the stars and stripes.[40]

If you are curious about the American robin's history at the dinner table, a recipe for robin pie can be found in *Wehmen's Cook Book*, which appeared in 1890 and was marketed as a collection of valuable recipes for every household:

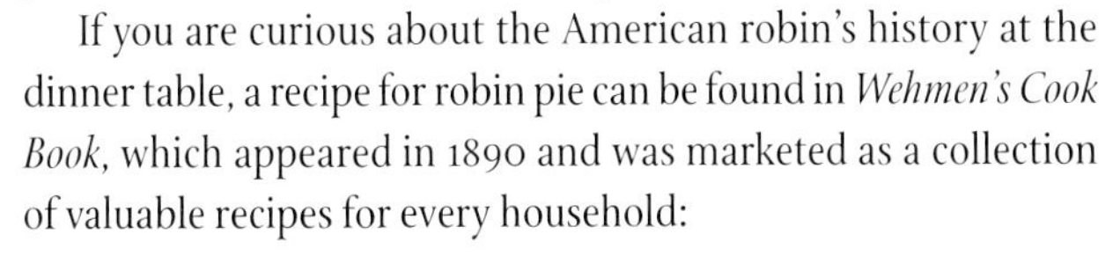

> Cover the bottom of a pie-dish with thin slices of beef and fat bacon, over which lay ten or twelve robins, previously rolled in flour, stuffed as above, season with a teaspoonful of salt, a quarter ditto of pepper, one of chopped parsley, and one of chopped eschalots, lay a bay-leaf over, add a gill of broth, and cover with three quarters of a pound of half puff taste, bake one hour in a moderate oven, shake well to make the gravy in the pie form a kind of sauce, and serve quite hot.[41]

Such a pie would be illegal in today's America, as the birds are now protected by the Migratory Bird Act of 1918, which also implements treaties that were entered into with Canada and Mexico.[42] In Canada, the bird is treated with similar affection as it is in the USA: a familiar bird and an icon of spring. American robins featured on the back of the Canadian $2 note, one of seven birds to appear as part of the Birds of Canada banknote series, which was first circulated in 1986. The note was withdrawn ten years later and replaced by the $2 coin. In 2013, as part of a celebration of the splendour of 'Canadian wildlife', the American robin also featured on the reverse side of a 25-cent coin, perched on an apple tree in blossom – a sight that Canadians in nearly every region of the country will be familiar with during the warmer months of the year.

Staying in the northern hemisphere, it is worth turning to the bird that was once described as the 'Eastern representative' of the European robin: the Oriental magpie-robin (*Copsychus saularis*).[43] Like the European robin, the Oriental magpie-robin is a member of the chat family and is placed in the genus *Copsychus*, along with a further four species of magpie-robins, all of which are noted for their ability to raise and spread their tail. Mainly ground-loving birds, they can be found across most of the Indian subcontinent and part of Southeast Asia, where they are common birds in forests and urban gardens.[44] Unlike the other robins covered so far, the Oriental magpie-robin doesn't have a brightly coloured breast, but in the eighteenth century it was nevertheless considered to resemble the European bird in both form and habit.

The Oriental magpie-robin offers an especially cautionary tale of European knowledge production. In 1867 the *Dictionary of Birds* documented a series of mistakes that had arisen as a consequence of the phonetic spelling of the bird's Indian name, which was *dhyal* or *dhayal*.[45] It was noted that the name had appeared in 1737 in Eleazar Albin's *A Supplement to the Natural History of Birds*, where it was referred to as the 'Bengall magpie' or 'Dial Bird'.[46]

Sketch of a bird in the blackbird family, 1843, ink and colour on paper.

This phonetic spelling of the Indian name led to some confusion. In the early nineteenth century, the French ornithologist François Levaillant made reference to Albin's description in his *Histoire naturelle des oiseaux d'Afrique*, in which he confessed that he did not understand the reason for the bird's name but supposed that 'Dial' was likely to be due to the bird's propensity for singing at certain times of the day – notably at sunrise or sunset.[47] Thus, he suggested, those familiar with the bird might use it to tell the hour.[48] Ironically, and to some amusement, Levaillant states that in order to avoid confusion he kept the name as Albin described, simply translating it as *oiseau-cadran*, using the French word for dial, as in sun-*dial* or the face of a watch or clock.[49]

It was thought by British ornithologist Thomas C. Jerdon that Linnaeus had also made a similar error when he assigned the bird its original Latin name, *Gracula saularis*. Jerdon assumed that *saularis* must have been a slip of the pen and should have been *solaris*, meaning 'of the sun'.[50] However, the suggestion that Linnaeus had made an error was contested by Edward Blyth, the curator of the Museum of the Asiatic Society at Calcutta.[51] *Saularis*

Doyel Chattar, Dhaka, Bangladesh.

was in fact a Latinization of the Hindi word *saulary*, which means a 'hundred songs'. This had been the name that Edward Bulkley – a surgeon with the East India Company – had sent from Madras, along with a specimen, to the naturalist James Petiver, who first described the species in the seventeenth century.[52]

While Thomas C. Jerdon had been wrong in his assessment of Linnaeus, in 1863 he noted that he had called the bird 'magpie-robin' in his own catalogue of 1839, and had since received information from the British diplomat, Mr Layard, that magpie-robin was the name by which it was now known by Europeans in Ceylon (now Sri Lanka).[53] Juvenile magpie-robins were said to resemble that of the European robin, and the bird's somewhat bold appearance (on account of carrying its tail very erect) was also said to be much like that of the English robin, according to the British naval officer Captain Boys.[54] It was these observed similarities, the bird's familiarity and the regard in which it was locally held that led it to be described as the 'Eastern representative' of Europe's own 'red-breasted favourite'.[55]

Oriental magpie-robins were frequently kept in cages for their song but were also kept for the purpose of fighting, owing to their 'pugnacious qualities'.[56] As one British naturalist put it, no race of game-cock could combat with more energy or resolution than the Dayal bird, thus providing much amusement for the wealthy of Nepal.[57] In Bangladesh, where the bird is known locally as *doyel* or *doel*, the magpie-robin is the national bird. Doyel Chattar (Square) in Dhaka is a tribute to the bird, which is located in front of Curzon Hall, a building from the British Raj-era that is now a part of the University of Dhaka. The Doyel Chattar sculpture depicts two magpie-robins with their wings outstretched and their tails characteristically erect. The national bird has also featured

Kitagawa Utamaro, *Great Tit, Japanese Robin*, c. 1790, polychrome woodblock print.

Ohara Shōson, *Robin on a Maple Branch*, 1935, colour woodblock print.

Japanese robin (*Larvivora akahige*), one of a few birds that could be mistaken for the European robin.

on the reverse side of the 2-taka banknote since it was first issued in December 1988.

The European robin has not always been the sole member of the genus *Erithacus*. Indeed, it was only recently that two other birds – the Japanese robin (*Larvivora akahige*) and the Ryukyu robin (*Larvivora komadori*) – were placed in the newly resurrected *Larvivora* genus after genetic testing revealed them to be dissimilar to the European bird.[58] The Japanese robin is noted to be one of few birds that could be mistaken for the European robin on account of its distinctive bright orange breast feathers, brown tail and olive-brown upper parts, although the extent of its orange plumage to its face and back distinguish it. The Ryukyu has a black face to breast and vivid rufous wings and upper parts. It is worth noting that while the Ryukyu robin's Latin name is *komadori,* this is the name that the Japanese robin goes by colloquially – a slightly perplexing detail!

Scientifically speaking, since the reclassification of these two East Asian birds, the European robin is apparently left with no close relatives. But, of course, in matters of culture the robin has many. In examining how the robin's name and associations travelled the globe, the extraordinary variety of other birds and histories that come into view are a reminder of the complex webs of relation that living beings are drawn into. The robin's global travels are an important reminder of the deeply cultural dimensions of so-called natural histories and the knowledge-making practices that have facilitated them.

3 Death, Bad Omens and a Pious Bird

For a small bird, the robin has a large presence in folklore. Robins are regularly cherished for their gentle companionship and familiar presence, but delve a little deeper and you find a far more complicated bird with a strong attachment to death. Robins are considered figures of charity and reverence, granting them an almost sacred standing, but they can also be terrible omens and bring disaster to one's door.

In cultural memory, the robin's name is often beloved thanks in no small part to the story of the children in the wood. From this tale comes the robin's reputation for having a charitable bill and offering dignity in death by covering dead bodies with leaves and moss. The first record of *The Children in the Wood* dates back to 1595, when it was published as a ballad by Thomas Millington in Norwich, England. Also known as *The Babes in the Wood*, or *The Norfolk Tragedy*, the tale concerns two children – a boy not older than three and his younger sister – who are left in the care of their uncle after the death of their parents. The children are to receive an inheritance, the boy when he comes of age and the girl upon her wedding day. The uncle, who has his eye on their inheritance, tells his wife that he is sending the children to be raised in London Town, but instead pays two 'ruffians' to murder them. Having lured the children into the woods, one of the men has a change of heart and is unable to carry out the mission. Instead, he kills the other man and takes the children a couple of miles into the woods where he leaves them and promises to return with food. The man is never seen again and the children die of exposure. While the story is miserable, a robin offers some comfort in the form of a burial:

Thus wandered these poor innocents,
Till death did end their grief;
In one another's arms they died,
As wanting due relief:
No burial this pretty pair
From any man receives
Till Robin Redbreast piously
Did cover them with leaves.

The wood depicted in the ballad is thought to be Wayland Wood in Norwich, and while there are multiple origin stories, the most credible is thought to have come from Reverend Charles Kent, who in 1910 suggested that the story had been Protestant propaganda based on a real-life event.[1] In the mid- to late 1500s, the Protestant owner of Griston Hall died, leaving behind a seven-year-old son to be raised by his uncle, a Catholic recusant.[2] When the son died at the age of eleven, the uncle tried to take the boy's inheritance, leading locals to accuse him of killing the boy for his money. Like many tales of its kind, the story of *The Children in the Wood* is a lesson in morality. The ballad continues:

And now the heavy wrath of God
Upon their uncle fell;
Yea, fearful fiends did haunt his house,
His conscience felt an hell:
His barns were fired, his goods consumed

'In one another's arms they died'. Robins fly in to cover the babes with leaves, illustration from *The Babes in the Wood* (1846–86).

His lands were barren made,
His cattle died within the field,
And nothing with him stayed.[3]

The ballad was reprinted in numerous chapbooks (a short book or pamphlet of poetry or short stories) and references to the robin's moral fortitude were to be found in other works, such as Michael Drayton's fable *The Owle* (1604): 'Covering with moss the dead's unclosed eye,/ The little redbreast teacheth charitie.'

Costume design by Attilio Comelli for the robins in *Babes in the Wood* at the Theatre Royal Drury Lane, 1907.

In 1851 the American sculptor Thomas Crawford (famous for the Statue of Freedom on top of the Capitol dome in Washington, DC) depicted *The Babes in the Wood* in a marble sculpture of the same title.[4] A robin watches over the children, whose untroubled, somewhat tranquil expressions paint a serene picture of death. The piece is considered a poignant example of Victorian attempts to 'soften the anguish of death by treating it with a bittersweet sentimentality'.[5] Child mortality was a common occurrence during the nineteenth century, so it is unsurprising that the story of *The Babes in the Wood* held particular poignancy and the pious robin redbreast became a source of comfort. This can been seen in 'The First Snowfall', written by the American romantic poet James Russell Lowell as a memorial to his first daughter. It includes the lines:

> I thought of a mound in sweet Auburn
> Where a little headstone stood;
> How the flakes were folding it gently,
> As did robins the babes in the wood.[6]

It is worth noting that both the European robin and the American robin are known to turn up moss in search of grubs, and that while *The Babes in the Wood* is a tale that has its origins in England, the same tale and references to piety appear in cultural histories of the American robin.[7]

The pious duties of the robin featured in William Wordsworth's *Memorials of a Tour on the Continent, 1820*, in which he documented

Thomas Crawford, *The Babes in the Wood*, 1851, marble.

his visit to the ruins of Forte di Fuentes, at the head of Lake Como in northern Italy. While the fort was largely ruinous, Wordsworth found glimpses of former splendour and was especially taken with a cherub that had survived relatively intact but had tumbled some way down a hill. The desolate cherub, as he described it, was sheltered in its place of rest by plants and shrubs, but when the winter came and the leaves fell, the poet imagined that the duty of kind shelter might be performed by birds as in the tale of *The Babes in the Wood*:[8]

> 'Dread hour! when upheaved by war's sulphurous blast,
> This sweet-visaged Cherub of Parian stone
> So far from the holy enclosure was cast,
> To couch in this thicket of brambles alone;
> . . .
> Where haply (kind service to Piety due!)
> When winter the grove of its mantle bereaves,
> Some Bird (like our own honoured Redbreast) may strew
> The desolate Slumberer with moss and with leaves.[9]

Shakespeare similarly referred to the charitable nature of the robin in *Cymbeline*. In a note to the text, which tells us something of the ballad's far-reaching significance, it is remarked that while the idea of robins covering the dead with leaves is surely an ancient idea, 'Shakespeare's readers would certainly think of the ballad of 1595, for in Elizabethan times it was the most famous of all ballads.'[10] Here, the robin performs a kind duty that others have neglected:

> the Ruddock would
> With charitable bill – O bill sore shaming
> Those rich-left heirs that let their fathers lie

Marcus Gheeraerts the Younger, *A Boy Aged Two*, 1608, oil on panel. Birds were often used to symbolize the human soul departing at the time of death.

Without a monument! – bring thee all this;
Yea, and furred moss besides, when flowers are none,
To winter-ground thy corse. (IV.2)

In 1695, a hundred years after the tale of *The Children in the Wood* was first published, the robin's connection with death was further cemented when a robin was seen to spend several days in Westminster Abbey while the body of Queen Mary II lay in state. The sight of the red-breasted bird perched atop the pinnacle of the queen's mausoleum caused something of a stir, prompting speculation about its meaning and generating a flurry of reports on what was dubbed the 'Westminster Wonder'. While the bird was occasionally seen to leave the abbey in order to find food, it remained close by and notably vocal. Some observers interpreted the robin's song as a mournful lament. Others feared that the bird's presence might be an indication of the queen's tortured soul, or, perhaps to the contrary, evidence that her soul had been safely received. Given the audacity of the bird to perch upon the mausoleum, some suggested that the robin was far more sinister: surely it was an enemy of the throne! Despite mixed interpretations, a published ballad struck a decidedly optimistic note:

A wise Astrologer declares,
It is a sign that our Affairs
Will be successful e'ery Spring,
Which makes the Robin Red-breast sing.
He learns from these sweet Songs of Joy,
That Potent France shall ne'er destroy
The Church, tho' good Queen Mary dy'[d]
For God above will be our Guide.[11]

There is a long history of robins choosing to spend their time in sacred places, which has provided ample fodder for local news, especially during colder months. In 1787 the *Bristol and Hotwell Guide* recounted the story of a robin who took up residence in Bristol Cathedral, where the bird was frequently heard singing during services, perched upon one of the pinnacles of the great organ. The robin was said to be so tame that it followed the vergers around in order to be fed and remained there until its death. One of the minor canons wrote a song in dedication, which contained the following lines:

> Now shake thy downy plumes, now gladlier pay
> Thy grateful tribute to each rising day;
> While crowds below their willing voices raise,
> To sing with holy zeal Jehovah's praise!
> Thou, perched-high shalt hear th' adoring throng,
> Catch the warm strains, and aid the sacred song!
> Increase the solemn chorus, and inspire
> Each tongue with music, and each heart with fire![12]

The superstitions that accompanied the tale are particularly noteworthy, for the robin appeared not long after the death of one of the cathedral's choristers, Edwin Everett, who was known for his exquisite voice. It was said that nobody in the congregation could have ignored the strange coincidence of the robin's arrival: the bird flew through the window, performed a circuit around the room and immediately joined the choir in song. Apparently it was 'as though little Edwin had come back to join the choir in the likeness of a Robin Redbreast'.[13] Like the Westminster Wonder, the robin of Bristol Cathedral offers a commentary on society's relationship with death, and there are many accounts like it. Robins appearing by a headstone or in the days immediately after a death

Cabot Tower, Bristol. There are many stories that concern robins and churches.

have been considered a sign from the afterlife, a visitation from a loved one or an indication that they are at peace.

Robins have also been known to bring great solace during illness or on the deathbed. In his poem 'The Redbreast', William Wordsworth described a cherished robin that comforted his sister during an illness that confined her to her bed. The robin had entered the house and taken up a position in her room, perched on a nail that had once supported a picture, from where he would sing and 'fan her face with its wings'. Such was the feeling that he brought that Wordsworth banished the cats from the house and declared the house a winter sanctuary for the bird: 'He needs not fear the season's rage,/ For the whole house is Robin's cage.'

Wordsworth's affection for robins was so great that when he too was taken ill and approaching the last years of his life, he published 'To A Redbreast', in which he asked the sweet bird to sing his requiem and forever herald spring:

> Then, little bird, this boon confer,
> Come, and my requiem sing,

Nor fail to be the harbinger
Of everlasting spring.[14]

It is on the basis of such varied tales of piety and compassion that there are so many horrifying stories of what might befall anyone who would dare harm a robin. News reports have included accounts of great illness, ruined crops, direct lightning strikes, cow's milk mixed with blood and the destruction of all of one's household crockery.[15] It has been said that in the Alpine regions of Austria and Italy, it was believed that anyone who killed a robin would suffer from epilepsy or the 'St Vitus dance'.[16] It is not clear what the origins of this belief are, nor at what point they emerged, but in *A 17th-Century Handbook of Bird-Care and Folklore* the Italian ornithologist Giovanni Pietro Olina described Robin Redbreast herself as having a tendency to suffer from dizziness or epilepsy.[17] In Ireland, small boys were told that if they robbed a robin's nest they would get sore hands, and in County Donegal the tale went that a robin would utter a curse if it found it had been robbed, which would go something like this:

Má's duine beag a thóg mo nead,
go dtabhairidh Dia ciall dó
Má's duine mór a thóg mo nead
Gogcuireadh Sé faoi chlár é

If it's a little person [a child] who took my nest
May God give him good sense
If it's a big person [an adult] who took my nest
May God send him to death[18]

In some superstitions, you needn't kill a robin for it to bring you or your loved ones harm, for the robin has also been considered a prophet of death. While the likes of Wordsworth welcomed the presence of a robin inside the house, for others a robin's short hop across the threshold was a bad omen, just as three taps on the window foretold a death.[19]

Whether feared or cherished, the depth of feeling for robins was abundantly clear when an account of a church parish in crisis was published in 1835. In *The Three Death-Cries of a Perishing Church*, the death of a robin was cited as evidence of a church in disarray. It was reported, with much incredulity, that a robin, who had been in the habit of attending St Mary's church in Nottingham, England, was ordered to be shot by none other than the archdeacon himself – and on a Sunday of all days. Such treatment of this poor churchgoer was described as 'revolting to all one's better feelings', on account of the robin being universally loved and widely considered sacred. If, as Wordsworth suggested, robin redbreast is the bird that man loves best, then it would appear that the archdeacon was the exception to that rule. As the publication insisted, even 'the very raggamuffin [*sic*] lads of the rudest village in England, who make a ruthless crusade against all other birds' nests, spare those of the robin and wren', because 'Robinets and Jenny Wren/ Are God Almighty's cocks and hens.'[20]

If the archdeacon's 'violation of a national and more than national attachment' to the robin wasn't enough to persuade the public of his unsuitable disposition, then further details of the story might. Upon taking aim at the bird, it was reported that a shot was fired into the house of one Mr Robinson, shattering eight sash-squares of glass. The archdeacon reportedly refused to pay for the damage.[21]

Such reports of an outrageous death lead us to another tale that might be considered as important to the robin's story as *The Children in the Wood*. The eighteenth-century rhyme 'Who Killed

Cock Robin?' continues to be one of the robin's most baffling cultural references. While seemingly an open-ended question, the rhyme leaves no doubt as to who is guilty of the crime, but quite what motivated the killing remains a mystery. The macabre rhyme first appeared in *Tommy Thumb's Pretty Song Book*, a chapbook produced in 1744, the first known collection of rhymes to be written down and published. Featured alongside familiar rhymes such as 'Sing a Song of Sixpence', 'Who did kill Cock Robbin' runs across three illustrated pages in black and red ink:

Who did kill Cock Robbin,
I said the Sparrow,
With my bow and Arrow,
And I did kill Cock Robbin.

Who did see him die,
I said the Fly,
With my little Eye,
And I did see him die.

And who did catch his blood,
I said the Fish,
With my little Dish,
And I did catch his blood.

And who did make his shroud,
I says the Beetle,
With my little Needle,
And I did make his shroud.

Verses were later added to cover Cock Robin's burial, which featured a cast of different birds who volunteered their assistance,

Cock Robin is shot with an arrow, from *The Death and Burial of Cock Robin* (c. 1830s).

including an owl who would dig his grave, a dove who would be chief mourner and a kite who would carry the coffin.

In the late eighteenth century, a companion rhyme to Cock Robin's death appeared in the form of 'The Marriage of Cock Robin and Jenny Wren'. The status of this companion piece remains in question, not least because this later addition offers something of a resolution to the original by depicting the killing as an accidental occurrence. Following the proposal of marriage and the subsequent wedding day – celebrated with cherry-pie, currant-wine and a fine concert – the worst happens:

When in came the Cuckoo,
And made a great rout;
He caught hold of Jenny,
And pulled her about.

The burial of Cock Robin, from *The National Nursery Book*.

SIGHING AND SOBBING FOR POOR COCK ROBIN.

Cock Robin was angry,
And so was the Sparrow,
Who now is preparing his
Bow and his arrow.
His aim then he took,
But he took it not right;
His skill it was bad,
Or he shot her in fright;
For the Cuckoo he missed,
But Cock Robin he killed!
And all the birds mourned
That his blood was so spilled.

Seventy-five years after Cock Robin appeared in *Tommy Thumb's Pretty Song Book*, the children's nursery rhyme became the basis for a piece of radical satire, which was published as a pamphlet by John Cahuac in the wake of the Peterloo Massacre of 1819.[22] The political use of the rhyme says a lot about the notoriety of Cock Robin's plight. The satirical piece protests the events that occurred in St Peter's Fields in Manchester, England, on 16 August 1819, when a peaceful mass-gathering of more than 100,000 people demanding parliamentary reform of representation was violently dispersed. This led to a massacre in which an estimated eighteen people were slain and many hundreds more were severely injured.[23] In this satirical piece, Cock Robin represents the men, women and children who lost their lives, while the sparrow represents the armed members of the yeomanry. In accompanying illustrations, men and women run for their lives, some clutching young infants to their chests, while others who have been knocked off their feet are trying to shield themselves from the swords of the yeomanry who bear down on them from their horses.

Cock Robin marries Jenny Wren on the cover of *Who Killed Cock Robin?* (*c.* 1905).

UNTEARABLE LINEN

Who killed Cock Robin?

Father Tuck's

"NURSERY Series".

Printed in Germany

In response to the all-important question – Who killed Cock Robin? – we hear at length from the 'bold sparrows', who brag about the killings and the honours that were subsequently bestowed upon them for crushing the call for reform. The magistrates who had overseen the charge were represented by ravens, who would just as well see the poor robins die than give them change, while further illustrations depict the government as a circle of birds, or 'THE BIG-WIG CREW', who were said to have decided, 'AT SIGHT, That SPARROWS might lawfully kill ROBINS outright'. The satirical piece sets out a prophecy:

These vile SPARROWS,
who kill ROBINS so callous,
Will, sooner or later,
Take their *swing* from a *Gallows*!

Beyond political satire, the tale of Cock Robin also provided the inspiration for one of the most significant examples of Victorian taxidermy: the tableau of *The Death and Burial of Cock Robin* (*c.* 1861). The tableau was Walter Potter's first major creation and took him seven years to build. It was said to have 'provided a constructive repository for large numbers of his spare stuffed birds' – a collection of birds that amounted to nearly one hundred in total![24] Cock Robin is depicted in a blue casket and the tableau remains faithful to the rhyme, save for the absence of a kite to carry him to his grave. *The Death and Burial of Cock Robin* is largely considered to have launched Potter's career, as he went on to develop a reputation for (sometimes grotesque, sometimes amusing) anthropomorphic arrangements, including the *Rabbits' Village School* (*c.* 1888) and *The Kittens' Wedding* (*c.* 1890). Indeed, Potter's museum was once described as 'a collection of objects so varied and unlikely that only seeing is believing'.[25]

Walter Potter, detail from *The Death and Burial of Cock Robin*, c. 1861, display case with 98 British bird specimens.

References to the peculiar death of Cock Robin also appear in painting. John Anster Fitzgerald (1819–1906) was known for his fairy paintings and in the 1860s produced several paintings of fairies capturing and killing robins, as well as taking their nests. While he created multiple paintings on this theme, they were never formerly considered a sequence and beyond the title of his piece *Who Killed Cock Robin*, the paintings don't appear to offer any further link to the nursery rhyme. As English scholar Nicola Brown suggests, it is perhaps the familiarity of the robins, which are pictured naturalistically, that make the struggle between life and death, birds and fairies, so discomfiting.[26] This is exacerbated by the grotesque depiction of the fairies who are far from the 'minute perfection of detail' and great beauty that fairies were

John Anster Fitzgerald, *The Death of Cock Robin*, 1860–70, watercolour on card.

ordinarily seen to present.[27] Fitzgerald's robins are surrounded, feasted upon and held captive in miniature worlds where the natural and supernatural coexist. It has been suggested that Fitzgerald's imaginative paintings were partially the product of an opioid addiction.

The detailed depictions of a grotesque fairydom might, in some part, be attributed to the hallucinogenic qualities of opioids, but it is also worth noting that, according to critics, Fitzgerald may have taken inspiration from Walter Potter's taxidermy masterpiece. While we can never know for sure whether Fitzgerald ever cast his eyes on Potter's tableau arrangement, there are certainly some similarities in form. This includes the way Fitzgerald presents death as a spectacle and mimics the lifelike arrangement of taxidermic display that was popular at the time.[28]

If Fitzgerald's fairy paintings are too ghoulish for you, then perhaps Disney's *Silly Symphony* from 1935 might be more to your taste. In this whimsical eight-minute piece, the tale of Cock Robin is recounted as an operetta with a somewhat surprising end. Struck by an arrow while serenading Jenny Wren, Cock Robin plummets from his tree, prompting the police to round up three possible murder suspects from a nearby drinking den. As a caricature of the Hollywood actress Mae West, Jenny Wren is a far cry from the blushing, plainly dressed wren of the original rhymes. With blonde hair, a feather ruff and a slow, provocative walk, Jenny Wren demands justice for her Robin during a dramatic trial. The parrot district attorney interrogates the three suspects and yells, 'There's the arrow, where's the bow, answer the questions yes or no,' while the low, deep voice of the owl judge serves to draw out the question, '*Whoooooooo* killed Cock Robin?' Outrageously, when a jury fails to identify the guilty party, all three birds are sentenced to hanging. Luckily, Cupid arrives just in time to reveal himself as the real culprit. Cock Robin isn't dead: 'He fell for little Jenny Wren and landed on his head.'[29]

Likened to a pre-noir crime story, the symphony was considered one of Disney's best, and is uncharacteristically satirical in its treatment of the U.S. criminal justice system.[30] Given his Hollywood makeover, it would be reasonable to assume that Cock Robin, with his red waistcoat and dark head-feathers, is an American robin. Indeed, despite the animators' insistence that there were no connections, Disney's Cock Robin was often likened to the hugely popular American singer and actor Bing Crosby, who was known for his crooning.

Glasgow coat of arms, 1866.

(*opposite page*) Untitled mural by Smug, High Street, Glasgow.

Disney's version of events had a far happier ending than the original tale, but there is another account of robins and death with an even more miraculous outcome. This concerns the miracle of St Kentigern (known as St Mungo) who, in the early part of the sixth century, was said to have brought a robin back to life after it had been killed. There are multiple variations on the story. In one, the robin is a companion to Mungo's teacher, St Serf, and is killed by a group of young men who are jealous of Mungo and try to frame him for the robin's murder. In another, St Serf's disciples kill the robin by accident, and in a further account it is a group of bored boys throwing stones at some ground-feeding birds. Whether bored boys or calculating young men, the offenders ran off. Mungo smoothed the robin's feathers and prayed until the lifeless bird awoke and flew away. The robin's revival was hailed as Mungo's first miracle, putting him on the path to sainthood.[31]

St Mungo went on to become the founder and patron saint of the Scottish city of Glasgow, and to this day the robin features in the city's coat of arms, along with the symbols of his other miracles – a tree, a bell and a fish. The miracles of St Mungo are also included in the University of Glasgow's coat of arms. In 2016 a mural of a modern-day St Mungo was completed by the artist

Theo van Hoytema, *Dead Robin*, *c.* 1895, lithograph.

TO LET

Smug, on High Street in Glasgow city centre. While the mural is officially untitled, the presence of two robins makes the identity of the subject immediately clear. Just around the corner on George Street is another untitled portrait of St Enoch and child, a tender portrait of St Mungo with his mother, who was credited with giving her son his love and compassion for nature. [32] As she cradles her son, a robin perches on St Enoch's wrist.

In drawing this preoccupation with untimely robin deaths and miraculous revivals to a close, it is probably apt to note that the natural mortality rate of robins is high. This is a fact that has not always been welcomed by those who have an especial attachment to them, as Ted R. Anderson noted in his biography of David Lack. According to a story that was still circulating twenty years later, Lack delivered a talk to a local bird club and was approached by an elderly woman, who questioned his findings on the high annual mortality rate of robins – which in the British Isles is thought to be around 62 per cent in adult birds. The woman insisted that she had been feeding the same robin in her garden for the last seventeen years. When Lack told her that this was highly unlikely, she began to hit him over the head with her umbrella.[33] While Lack was likely right, the oldest recorded robin on record is in fact seventeen years and three months, although very few ringed birds are recovered after four years.[34] It is thought that the average robin is likely to live for six years.

4 A Bird of Song and Seasonal Change

Birdsong has long been a source of artistic and literary inspiration. After reading Rachel Carson's *Silent Spring*, in which she details the environmental devastation wrought by pesticides, the philosopher Aldous Huxley was purported to have exclaimed, 'we are losing half the subject-matter of English poetry.'[1] The robin's melodious repertoires have provided rich focus of attention for poets, writers and artists alike. Known for its beautiful song and aesthetic form, the robin has earned itself a reputation as both a melancholic figure and a joyous symbol of return: a harbinger of seasonal change.

For a description of the robin's song, it would be difficult to better the one provided by the British ornithologist Peter Clement, who offered the following notes:

> A melodious, wistful, lilting series of up to three, fairly slow or unhurried, liquid, musical, high and low whistles, and twittering notes, given in phrases broken by short pauses, with each series varying slightly from [the] previous one . . . individual singers have extensive repertoires, overall fairly mellow, wistful or mournful in quality especially on autumn or winter evenings, when usually the only bird singing.[2]

Theo van Hoytema, *Robin on a Snowy Branch*, 1878–1910.

The robin's song in autumn is said to differ slightly from that of spring, containing 'softer, longer and mellower phrases' which might be considered to possess an almost 'sad quality'.[3] Such perceived difference between the two songs – autumn and spring – has been frequently discussed: might it be that the difference says more about the mood of the listener than anything else? In short, how you hear the robin's song might be shaped by how you

A robin sings from a vantage point.

feel about the seasons. In his popular *The Charm of Birds*, Edward Grey, 1st Viscount Grey of Fallodon, noted the robin's autumn song to be somewhat 'thin', but wondered whether this is because 'our own minds are attuned to a minor key'. Maybe, he suggested, if one were to listen to a robin in the glow of spring and detect a more joyful note, one should ask, 'Is it the song or is it I that have changed?'[4] While Grey's musings on the robin's song are well documented, the bird was said to have been given a dreadful character reference from the viscount, who described the robin's personal life as a disgrace on account of its aggressive pursuit of territory. Perhaps, he supposed, a bird of such fighting spirit will always find favour with humans who are conscious of their own struggle for existence.[5]

In England, the robin's appearance in poetry coincided with the improvements in domestic conditions seen from the eighteenth century onwards. Before such improvements, the arrival of autumn was a reminder that the cruel and deadly months of winter were around the corner – a rather bleak horizon and certainly nothing to celebrate! But as domestic comforts improved, autumn was no longer a season to dread and was to become the focus for some of the greatest poetry of the Romantics.[6]

In wanting to capture something of the feeling of autumn with which robins have been associated, it is worth quoting from John Keats's 'To Autumn':

Individual robins have extensive repertoires.

Where are the songs of spring? Ay, Where are they?
Think not of them, thou hast thy music too,—
While barred clouds bloom the soft-dying day,
And touch the stubble-plains with rosy hue;
Then in a wailful choir the small gnats mourn
Among the river sallows, borne aloft
Or sinking as the light wind lives or dies;
And full-grown lambs loud bleat from hilly bourn;
Hedge-crickets sing; and now with treble soft
The red-breast whistles from a garden-croft;
And gathering swallows twitter in the skies.

Keats's biographer, Andrew Motion, describes 'To Autumn' as a pastoral poem that balances life and death. The poem, he suggests, has a 'feeling of penultimacy'. The processes Keats describes – the fruit ripened to the core, the last oozings of the cedar press, the soft-dying day – are always seemingly on the cusp of ending, but not quite. This is keenly felt in the last four lines of the poem, where the robin makes its appearance. The swallows that gather but are not yet departed are a reminder of the absence that is soon to come. As the robin's song begins, we sense that a turning point is around the corner and winter is on the way; the 'penultimacy' that Keats depicts cannot last.[7]

In possessing a somewhat mournful quality, the robin's song lends itself to melancholy, a state with which the robin has been frequently associated in natural and cultural histories alike. Melancholy is described as a feeling of pensive sadness. As English literature scholars Martin Middeke and Christina Wald acknowledge, the condition of melancholia has often enjoyed a rather noble cultural status: while it is a painful state, it is believed to open up 'an avenue to deeper insight, to judiciousness and to creativity'.[8] In this vein, the wistful notes of the robin have been said to inspire contemplation, and it is for this reason that Wordsworth described the robin as the 'pensive warbler of the ruddy breast'.[9] Despite its rather sombre tone, melancholy is complex and connected to joy, for it is said that melancholy can only be known by those who are capable of experiencing joy's loss.[10] And so, upon hearing the redbreast's autumn whistles, it is said that we are keenly aware of the joy that is soon to pass.

Keats's robin might strike a melancholy tone, but for Wordsworth the passage of time and the robin's song could also be rather rousing. After writing very little in the summer months of 1804, Wordsworth described a red-breasted intervention that stirred him to activity and announced the onset of colder months:

> But I heard
> After the hour of sunset yester-even,
> Sitting within doors betwixt light and dark,
> A voice that stirred me. 'Twas a little band,
> A quire of redbreasts gathered somewhere near
> My threshold, minstrels from the distant woods
> And dells, sent in by Winter to bespeak
> For the old man a welcome, to announce
> With preparation artful and benign –
> Yea, the most gentle music of the year –
> That their rough lord had left the surly north,
> And hath begun his journey.[11]

Sentimental attachment to robins is not the preserve of poets but can also be found in the paintings of J.M.W. Turner, who was working at the same time as Keats and Wordsworth. It may come as a surprise that Turner, the painter best known for his extraordinary landscapes and turbulent seas, painted not only a robin but a whole portfolio of birds. In his account of *Turner's Birds*, David

Hill suggests that 'anyone seeking to discover the real Turner could hardly start at a worse place than his exhibited oil paintings,' for they reveal only his public face.[12] Instead, Hill suggests that it is in Turner's bird watercolours that the real man can be discerned.

Between the years 1808 and 1824, the artist was frequently entertained at Farnley Hall in Yorkshire by his friend Walter Fawkes, with whom he shared a great interest in ornithology. During his time there, Turner contributed many illustrations for various albums, including a five-volume ornithological collection, which, alongside drawings, contained feather specimens and the occasional beak. The robin was one of twenty drawings of birds that he contributed, the majority of which were found in the local area. The robin, unlike most of the birds depicted, was painted from life and is noted to be decidedly weaker in execution and observation as a result, Turner being reliant on only fleeting glimpses of the bird.[13] Turner's bird paintings were not considered to be

J.M.W Turner, *Robin Redbreast*, c. 1815, watercolour.

The American robin, harbinger of spring.

the most accurate or scientific by any means, but they have been noted for capturing a sense of marvel and a quality of intimacy – a delight of close scrutiny and a concern for making visible the feelings that birds inspire.

The robin that inspired Turner's expressive colourization might be strongly associated with autumn and the imminent arrival of winter, but the American robin – the quintessential early bird – has a much stronger association with spring and its own poetic archive as seen in William Warner Caldwell's 'Robin's Come':

> From the elm-tree's topmost bough,
> Hark! the robin's early song,
> Telling, one and all, that now
> Merry spring-time hastes along;
> Welcome tidings thou dost bring,
> Little harbinger of spring!
> Robin's come.[14]

The robin's connection with spring is in evidence elsewhere. Take the wake-robin, the name given to a variety of North American spring flowers that are indicative of the arrival of the season.

The American naturalist John Burroughs wrote: 'When I have found the wake-robin in bloom I know the season is fairly inaugurated. With me this flower is associated, not merely with the awakening of the Robin, for he is awake some weeks, but with the universal awakening and rehabilitation of nature.'[15] In this matter, it is not only spring flowers that take the robin's name. The phrase 'robin snow', which was used in New England and has its origins in the mid-nineteenth century, refers to a light snowfall that occurs after the arrival of the first robin, or a snow that is light enough that it doesn't drive the robins off.[16]

Many American robins will travel south for the winter and can be found from Central Mexico through to southern Canada, but there are many across North America that remain resident throughout the year and will spend the winter feeding on berries. Over the years, the annual berry feast has been the source of some amusement. In the notes section of the 1930 edition of the *The Auk*, an incidence of intoxicated robins in Denver, Colorado, was reported. Over several seasons it was observed that after greedily eating an abundant crop of red berries from a honeysuckle bush, robins were seen to suffer from what was described as a condition of more or less profound intoxication, resulting in anything from a mild unsteadiness to such ill coordination that they fell to the ground.[17] In her stories of the 'urban wild', the writer Lyanda Lynn Haupt described tipsy robins that will sit on back porches in such drunken stupors that you can walk right over and pick them up.[18] American robins in northern Minnesota even made the national news in 2018 when there was a spate of reports of 'birds flying under the influence'. The local sheriff confirmed that a number of birds had been taken to the station and given the chance to sleep it off and sober up.

There may be numerous anecdotal reports of 'drunk' backyard birds across North America, but while it is known that alcohol

An American robin among autumn berries.

forms in berries as they ferment with the first frosts, there have been few studies to confirm that fermented berries are the cause of this 'drunken behaviour'. As one scientist put it, there is no routine test for diagnosing alcohol poisoning in animals. However, this propensity to overindulge, even if anecdotal, is something that it shares with the European robin. According to *A 17th-Century Handbook of Bird-Care and Folklore*, it was well known that the robin had a taste for gooseberries, which were found to occasionally make it merry.[19]

Alongside the North American and European birds, there is another robin that is associated with the passing of the seasons. However, the bird formerly known as the clay-coloured robin is now more likely to go by another name: the clay-coloured thrush (*Turdus grayi*). This common bird – the national bird of Costa Rica – has a range that covers eastern Mexico through to northern Colombia. Known locally as the *yigüirro* (from the Huetar language of Costa Rica), the bird has much in common with the American robin in that it is alike in form and habit. It was for this reason that it was originally given the robin name – yet another demonstration of how vernacular names not only travelled but were imposed on birds the world over. In Costa Rica, when the much-needed rainy or 'green' season arrives, this plain grey-brown bird of national adoration resumes its song and is said to 'call the rains'. When the rains are late to arrive, the bird is implored to sing.[20] According to the Bribri people of Costa Rica, it was through song that the clay-coloured thrush contributed to the dance of the Creation.[21]

Beyond their ability to call the seasons or sound their passing, it is their melodious strains that have made songbirds of all kinds popular pets. While in most of their range it is now illegal to keep European robins in a cage, this hasn't always been the case. The caging of a robin was the focus of a nineteenth-century Dutch

Clay-coloured thrush (*Turdus grayi*), Costa Rica.

fairy tale for children. Written by Reinoudina de Goeje under her pen name, Agatha, *The Robins* charts the adventures of a robin who tries to track down his mate after she was captured for her song and sold to a king and queen.[22] Despite encountering numerous dangers along the way, the robin persists and finds his mate in a cage at the palace. When reunited, the two sing so strikingly together that they draw the attention of the king and all the lords and ladies of the court, who gather to listen to the mournful song. Having declared that she understands the birds, the queen orders the cage to be opened and predicts that the song of the caged robin will no longer sound so sorrowful but will instead be full of cheer. The queen was right and the birds fly away to build a new nest that is entirely hidden from view, where they sing from morning to late: let us free; let us free! *Want slechts in vrijheid, is voor ons blijheid.*[23] ('For only in freedom, is for us happiness.')

As is perhaps evident from the story's moralistic elements (for which Agatha was widely known), even while the capture

The queen sets the robin free, illustration from *The Robins*, by Agatha (1874).

of robins was once not so unusual, it was not a practice that was universally embraced and the morality of it was sometimes debated.[24] When it was embraced, robins were just as likely destined for the pot as taken for their melodious qualities on account of being considered 'a light and good meat'.[25] Such enthusiasm for their meat was documented by a number of prominent naturalists, the French natural historian the Comte de Buffon included. (You only need to look at seventeenth-century Dutch still-life paintings to know that robins were a regular at the dinner table.) While there was variability across the European continent, a gradual shift in popular sentiment meant that the caging of small birds steadily died out as a practice. In fact, in parts of England, even when times were hard, not everybody was at ease with capturing robins in the first place. Documents suggest that despite the scarcity of food, the accidental capture of a robin would be a sad occasion and the poor bird would be promptly buried.[26] This says much about the cultural status of the bird and its various superstitions. As William Blake's famous line goes: 'A robin redbreast in a cage/ Puts all Heaven in a rage.'

The European robin is no longer legally caged for its meat or song, but this is not the case for other robins. Take the Oriental magpie-robin, whose slow and tuneful whistles have made it a popular cage bird in Southeast Asia and whose numbers have been historically impacted as a result. In Singapore, for example, the Oriental magpie-robin was reduced to single unpaired individuals by the early 1980s.[27] While their numbers have since recovered as a result of increased protections, the Oriental magpie-robin still routinely features on bird lists associated with wildlife trafficking in Southeast Asia as well as the lucrative songbird trade. In market surveys in Indonesia, Oriental magpie-robins were not only identified as one of the species most traded and trapped but were also shown to be increasing in price and value and

C.S.S.

were flagged as being of especial concern as a result, despite stable numbers internationally. In 2016 they were identified as one of 28 priority species deemed to be threatened by songbird trade in the Greater Sunda region, comprising the islands of Java, Sumatra, Borneo and Sulawesi.[28]

Within Southeast Asia, the demand for songbirds as pets and for use in songbird competitions involves hundreds of species and millions of individual birds annually.[29] In response to what has been dubbed 'the Asian songbird crisis', the IUCN SSC Asian Songbird Trade Specialist Group (ASTSG) was launched in 2017 with the aim of preventing the extinction of songbirds threatened by illegal and/or unsustainable trade. It is an issue with complex social and cultural dimensions.[30] For instance, in Indonesia, bird keeping has deep and diverse cultural roots and is an integral part of everyday life. Birds are considered a source of companionship, care and aesthetic pleasure and have thus come to prominence in contemporary urban culture and homelife, as well as being an important means of retaining a connection to rural village life.

Songbird competitions have enormous popular appeal, and appreciation for birds is often founded on a deep and sophisticated knowledge of birdsong, behaviour and husbandry.[31] Started in the 1970s by a group of bird enthusiasts in Jakarta who waged significant prizes on the outcomes, bird-singing contests have become a major hobby, giving rise to the phenomenon 'kicau mania'. Birds are judged on melody, duration and volume, with sizeable cash prizes on offer. While earlier competitions initially focused on imported Chinese birds, native species were eventually incorporated and came to replace the imported birds as contest popularity grew. In a typical contest, birds in ornate cages are hung on a metal frame some way off the ground and a metre away from fellow competitors. The birds are encouraged to sing by their supporters with whistles and waves, while judges walk around

Cornelis Samuel Stortenbeker, *A Great Tit and a Robin*, c. 1860–85, oil on panel.

Bird-singing contest, Thailand.

and assess their repertoire.[32] Birdsong is talked about in ways akin to music, and birds are referred to 'as individual exponents of a score'.[33] The Oriental magpie-robin has an official song-contest class.

Robins have not only been appreciated for their song, but their refrain has been the subject of several hits. For this we leave South-east Asia and head to the USA for Al Jolson's 1926 upbeat release 'When the Red Red Robin (Comes Bob, Bob, Bobbin' Along)'. The song was about the American robin and has been credited with giving the popular Red Robin burger chain its name. The founder of the North American restaurant chain had been part of a barber-shop quartet that had regularly sung the hit. He was said to enjoy the song so much that his original restaurant, Sam's Tavern, was renamed Sam's Red Robin. When the restaurant was bought and expanded, 'Sam's' was dropped from its name.

The American robin was also the focus of Bobby Day's hit song 'Rockin' Robin', in which a robin was taught to bop by a raven and was loved by all the other birds for its song. Curiously, the American robin was absent from the sleeve cover of the original release, which featured all manner of birds, including a parrot. If streaming the song today, you might also find artwork that features the European robin rather than the American bird that inspired the song. Bobby Day's song was famously covered by Michael Jackson in 1972 but was also the basis for an advert in the UK for Fox's Rocky biscuit bar in the 1990s.

While the Bobby Day song was about the American robin, the Fox's advert depicted a European robin in a leather jacket, with his name emblazoned on the back. The popular Rocky Robin advert was given a revamp in 2003 to promote the new Rocky Rounds chocolate biscuits. The new adverts featured an almost spherical robin by the name of Rocky R who had swapped the rock 'n' roll of earlier adverts for rap. Wearing a purple jacket and gold medallion, Rocky R is seen to be living the life of a rap star while gorging on Rocky Rounds. It was unclear what prompted Rocky Robin's makeover and my enquiries have gone unanswered. Whatever the reason, unlike Bobby Day's song, Rocky R didn't last long.

In 1964 cultural references were also famously crossed during the song 'A Spoonful of Sugar' in Walt Disney's *Mary Poppins*. As the super-nanny tries to convince the young Banks children, Michael and Jane, that it is possible to have fun while tidying their room and doing their chores, she tells them to do as the robin does while tending to its nest: sing a merry tune. As she sings, Mary Poppins leans out of the bedroom window on Cherry Tree Lane where we see a pair of robins building a nest. She encourages one to perch on her hand and accompany her in song. Despite being set in London, with questionable cockney accents to boot, the observant viewer might be surprised to see that it is an American

John Baeder, *Red Robin*, 1982, screenprint.

robin that has set up home outside the Banks's family residence and not the European bird as one might expect.

Those concerned about this error will be pleased to know that it was corrected in the recent *Mary Poppins Returns* (2018), although they may be perturbed by the manner in which the robin appears – pinned to Mary's hat, with wing feathers fanned out. At the height of the so-called 'plume boom' in the early part of the twentieth century, such display was popular in ladies' fashion, and the wings and bodies of birds were frequently used for hat adornments (although the use of robins was uncommon).[34] It was on account of this most recent role that the robin made its first appearance on the front page of the women's fashion magazine *Harper's Bazaar*. In the interview with the film's star, Emily Blunt, it was suggested that the stuffed bird might perhaps serve as a useful device for preparing viewers for a 'less saccharine' interpretation of the childhood classic.[35] Readers can be assured that the avian star of the photoshoot was very much alive and went by the name of George.

Mary Poppins (Julie Andrews) with an American visitor, in *Mary Poppins* (1964; dir. Robert Stevenson).

Mary Poppins is not the only film in which the appearance of an American robin has been questioned. Writing in the journal *Audubon*, Peter Cashwell noted similar consternation when the bird made a brief appearance in *The Hobbit: An Unexpected Journey* (2012), the first of Peter Jackson's second trilogy set in the world of J.R.R. Tolkien's Middle-earth. The robin stars in a scene with the forest-dwelling wizard Radagast as he surveys the poisoned forest and wonders aloud where the dark magic is coming from. The robin appears by his side and offers to lead the way, hovering long enough to be easily identified. But as Cashwell points out, while Tolkien was known to have taken inspiration from the English countryside (and the scene features many European hedgehogs), the country's 'distinct lack of towering crags, impenetrable forests, and erupting volcanos should make clear, England is not Middle-Earth'.[36] His point was straightforward: no one can say where an American robin should or shouldn't belong in a world of make-believe.

The conflation of the two species is not restricted to song and film. The popularity of the colour 'robin's egg blue' has also led to some confusion. Known for its soothing, tranquil properties, this particular shade of turquoise is based on the egg of the American robin.[37] Because the popularity of the colour extends far beyond North America, it has occasionally led to some puzzlement over the identification of European robin eggs. Studies into the colour of American robin eggs have found that the bluer the egg, the healthier the mother is and the more time and energy the male robin will invest in feeding the chicks. The blue colour arises from a concentration of the pigment biliverdin, which in humans is responsible for the green colour sometimes seen in bruises.[38]

Robin's egg blue brings us to a different season: breeding season. Unlike its American counterpart, the egg of the European robin is around 2 centimetres (¾ in.) in length and is a mostly whitish colour with red speckles.[39] The breeding season for the European robin starts in the spring, with many pairs attempting two or three broods. On average, robins lay four or five eggs, with the most that any one nest might contain being limited to about seven. An egg is laid each day and they generally hatch after thirteen days of incubation, which only begins once all of the eggs have been laid. The incubation of the eggs and tiny young is largely carried out by the female, who is fed by the male. Once the young are a little older, both parent birds are engaged in the task of keeping hunger at bay. Like a number of other birds, they keep the nest clean and concealed from predators by removing the droppings of their nestlings, which are produced in little membrane sacks that can be easily carried away from the nest or even eaten. The eggshells are also taken away from the nest and the young birds fledge at about fourteen days old.

The robin's melodious song might have gained artistic and literary fame as a harbinger of seasonal change, but robins have also earned themselves a reputation for their unusual choice of nest sites. A robin's nest is often cup-shaped, made out of dry grasses, moss, dead leaves and other plant fibres, and further lined by finer materials such as feathers or animal hair. Natural sites are often found in hollows or nooks, either in trees or where the nest can be concealed – perhaps by ivy or another form of climbing plant. They are often built at a height of no more than 5 metres (16 ft) above ground but they can also be found in the crevices of roots, rocks or earth banks.[40] While there is nothing unusual here, it is the robin's propensity for using human-made locations that has earned it a somewhat comedic reputation. Holes in walls with a concealed entrance, disused vessels such as plant pots or kettles, the nooks and crannies of outhouses, as well as artificial nesting boxes, are common sites. But then there are the more outlandish:

American robin eggs or 'robin's egg blue'.

European robin eggs are whitish in colour with red speckles.

the engine of an RAF de Havilland Mosquito aircraft, the coat pocket of a hanging jacket, the folds of an unmade bed, garden boots, church lecterns and jam jars have all been reportedly used.[41]

In a frequently cited example, a robin was found to have built her nest in the skull of a convict who had been hanged in 1796. After being sentenced for assault and stealing a mailbag from the Warrington mail, James Price and Thomas Brown were hanged at Trafford Green in the northwest of England. When their bodies were finally removed in 1820, a robin's nest was discovered in the skull of James Price. This little rhyme was published in *Ballads and Legends of Cheshire* (1867):

> Oh! James Price deserved his fate:
> Naught but *Robbing* in his pate
> Whilst alive, and now he's dead

Robins make good use of an ornamental teapot.

Has still *Robin* in his head.
High he swings for *Robbing* the *Mail*,
But his brain of *Robin female*
Still is quite full; though out of breath,
The passion e'en survives his death.[42]

Unusual instances of nest-building have been of equal interest to natural historians. In his *Natural History of Ireland*, William Thompson recounted several examples during the summer of 1833. This included a robin, which, after having built an unsuccessful nest between a number of pickle jars on a pantry shelf, tried her luck in the flat upstairs (this was perhaps a bold move, for the flat was home to a 'bird-stuffing' enterprise). After attempts to dissuade the bird from entering, including the placement of a most 'fierce-looking' specimen at the window, the bird's perseverance was rewarded with the freedom to come and go as she wished. In what was taken to be an act of defiance, the bird eventually built her nest on the head of a shark that was mounted on the wall. During the months that followed, man and bird worked side by side, and while the man presumably continued with his taxidermy work, the robin successfully incubated five eggs.[43]

The unusual nest sites of robins have provided ample fodder for local newspapers, the archives of which are full of amusing observations and noteworthy accounts. These include nests built on the edge of a parlour bookcase; between a stove pipe and a wall in a corporation piggery, a crematorium (they were apparently undisturbed by the heat), the letter box on the entrance gates to a bowling club, an engineering bench, the dashboard compartment of a car (they went on multiple rides as a result), on a pile of books in a school cupboard and in a workman's haversack that had been hung up for the weekend, to name just a few.[44]

It is perhaps the speed at which a robin can build a nest that gives rise to so many of these accounts. Leave a window open or your coat hung up on a peg – even your bed unmade – and a robin can build a nest within hours. In most cases, however, this is probably a little too fast. It is ordinarily only the female that undertakes the task, and while robins can be fast nest builders, they are more likely to build a nest in shifts across a couple of days.

The location of robin nests is not the only thing that is worthy of note (which undoubtedly contributes to the wider affection for the bird), but the lengths that people will go to accommodate their unexpected guests: windows left open for ease of access, new sleeping arrangements, worms left in school cupboards or the temporary abandonment of church lecterns. If giving up your bed for a bird seems a little much, then you might spare a thought for the residents of a small village in Tamil Nadu's Sivaganga district in India, who in 2020 went without street lighting for nearly forty days in order to accommodate a nesting Oriental magpie-robin

When it comes to feeding the young, both parents take up the task.

that had laid her eggs in a switchboard box. It would seem that the European robin is not the only robin to seek out unusual sites. While the Oriental magpie-robin, which is widespread throughout India, will often build their nests in crevices or holes in tree trunks, they are also known to readily accept artificial sites.[45]

5 For Territory and Nation

Birds play an important role in national psyche, and the robin is no exception, but what does it take for a bird to become a national treasure?

National identity can be developed in rather banal ways. Take the postage stamp. These small pieces of paper carry a wealth of meaning and have a value far beyond their initial purpose. For the philosopher Walter Benjamin, stamps were enchanting miniature versions of the world and something to marvel at. In Benjamin's eyes, an album of stamps was a fantastical reference book which enabled viewers to travel to other places and learn of different cultures, flora and fauna without the need or ability to physically move.[1] If stamps offer a way to learn about other people and places at a distance, then they also play an important role in the development of a national imaginary, communicating something of a nation's achievements, history and core values. Romfilatelia, the institution responsible for issuing Romanian stamps, captures this role beautifully when describing the Romanian stamp as an 'ambassador' for the country, one that carries the 'treasures' of Romania wherever it goes.[2]

When it comes to stamps, birds are a popular subject and the robin has featured on stamps the world over. Whether in celebrating native fauna and flora, championing conservation and nature protection programmes, or showcasing species from

European robin stamp, Europa Protection of the Environment series, Lichtenstein, 1986.

elsewhere, the robin has been carried around the world with many an 'ambassador', from Andorra to Sierra Leone. Known as the 'red craw' in Romania, the European robin featured in the 2016 series promoting the Ceahlau National Park, while in Malta the robin appeared in front of a Maltese fortification to mark the 25th anniversary of the Maltese Ornithological Society. It has appeared alongside snowdrops in Syria, Beethoven in Bosnia and Herzegovina and as part of a bird protection drive in Monaco. In Ukraine, the robin was part of a wider celebration of the seasons, for which it appeared in the spring series, which showcased the generosity of the season and the country's abundance of native flowers. The robin could be found perched atop a handful of snowdrop windflowers.

If we can learn something from the styles and images of national stamps, then stamp collectors also point to their ability to tell us something of countries and regimes that no longer exist or have gained independence.[3] The robin appeared on stamps

Robin on windflowers stamp, Ukraine, 2011.

for Czechoslovakia (1964), Yugoslavia (1987) and the German Democratic Republic (1979), in each case as part of a celebration of native species and a drive to protect nature, but now a part of the material residue of history that has outlived the community it once represented.

Postage stamps are not the only miniatures to showcase national treasures. In 1879 the cardboard stiffeners in U.S. cigarette packets were replaced with colourful, collectable cards to foster brand loyalty in a highly competitive tobacco market (without the cardboard inserts, cigarettes were liable to sustain damage due to the flimsy nature of the paper wrapping). Mostly designed to appeal to men, cards were both visually appealing and informative and were intended to offer 'a window onto the wider world'.[4] Along with sports and the military, birds were a popular topic, with some of the earliest British sets featuring the robin. In Ogden's 1909 British Birds series, the robin was described as

> perhaps the bird dearest to all Englishmen, whose attributes he, in some measure, has adopted. He is not too sociable, his own company being quite sufficient, and he is not lacking in readiness of resource, while he evidently regards the presence of himself as ample recompense for a liberal board. The manner in which he will monopolise any locality he has selected, and drive away other birds, is most amusing.[5]

Birds of America series of cigarette cards to promote Allen & Ginter cigarettes, 1888.

Like the postage stamp, the cards helped to build a national imaginary, with the resourceful, self-reliant robin being imbued with a certain sense of national character and fortitude. In Britain, it was not only tobacco companies that issued collectable cards. When the rationing of tea ended in 1952, and price controls were abolished after a steady improvement of tea supplies following the end of the Second World War, tea companies were required to compete in an ever-growing consumer market. As a response, Brooke Bond introduced picture cards as part of its new advertising campaign. For their inaugural set they enlisted the renowned naturalist Frances Pitt to produce a twenty-card series of British birds.[6] The robin and its 'charming impertinence' featured as number sixteen: 'There are few birds more familiar or loved then the Robin or Redbreast. Its song delights us nearly all the year round.' The Brooke Bond cards were so successful that they established a department for the sole purpose of answering card requests.[7]

In the case of Ogden's, the robin wasn't to be confined to the insides of a cigarette packet, for the bird 'dearest to all Englishmen' became the face of two of Ogden's most iconic products: 'Robin Cigarettes' and 'Redbreast Flakes'. Widely popular in Britain, the success of these products also saw the robin feature in advertisements and supermarkets on the other side of the world. And Ogden's were not the only Victorian company to do well with a 'robin brand'. In 1890, Reckitt & Sons (later to become Reckitt & Colman) launched Robin Starch, which similarly benefited from a booming export trade thanks to the reach of the British Empire. Robin Starch was to become one of the best-known brands for stiffening collars and 'lending wings to your iron'. The trademark robin was said to be an assurance of quality.

Stamps, cigarette cards and other merchandise might play a rather banal role in shaping national imaginaries, offering a window onto the wider world and revealing something of what societies hold dear, but there are also more formal mechanisms for establishing national sentiment. In 2014, David Lindo, otherwise known as the Urban Birder, launched a campaign to name Britain's favourite bird, having noted that Britain was lacking an official one. Following a preliminary vote, which cut a long list of sixty

Robin cigarettes.

Reckitt & Sons Robin starch, 19th century.

birds to a shortlist of ten, the robin went on to win the public vote in 2016, with 34 per cent. With more than 200,000 votes, the robin scored nearly three times the number of votes received by the runner-up: the barn owl.[8] That the robin won this popular vote was perhaps unsurprising, given its familiarity and long-standing position in Britain's cultural heritage. It was a position that was further confirmed by a children's vote, which took place on the last day of the two-month campaign. The robin had already been named Britain's national bird in 1960 by *The Times* in what was, admittedly, a much less democratic vote. Yet while the robin has twice been granted the title of Britain's favourite bird, it is not a status that has been officially recognized and Britain is still without an avian representative.

The robin may not be an official emblem, but it continues to enjoy a certain national status, nonetheless. In 2018 the Royal

The R coin from the UK Royal Mint's 'Quintessentially British A-Z' 2018 collection of 10p coins.

Mint released 26 10-pence coins as part of the 'Quintessentially British A to Z' collection in which the robin was to be found nestled between 'Queueing' and 'Stonehenge'. The Loch Ness Monster aside, the nation's so-called favourite bird was the only creature to feature in this celebration of 'everything British'. The robin's inclusion in this celebration of national identity saw it up there with cricket, the Houses of Parliament, fish and chips, and tea. Not an insignificant feat for such a small bird! But what does it mean for a bird to be a national icon and why does a nation need a favourite bird in the first place?

There are many nations with an official bird, each with very different reasons for their selection. For instance, the bald eagle has been a national emblem of the United States since 1782 because it is considered to be a symbol of strength, although Benjamin Franklin famously described it as a bird with bad moral character.[9] The white stork is the national bird of Lithuania because it is considered a symbol of harmony, while in Bangladesh it is the Oriental magpie-robin that has the title because of its familiarity

Sheikh Zain ud-Din, *Oriental Magpie-Robin with Katydid and Leaf Hopper on Monkey Jack Branch*, 1778, opaque colours and ink on paper.

throughout the country. It is not only nations that have official birds. While the bald eagle might be the national emblem of the USA, the American robin was adopted as the state bird for Michigan in 1931 after nearly 200,000 votes were cast in an election organized by the Michigan Audubon Society to find the state's most popular bird. It was concluded that 'the robin redbreast is the best known and best loved of all the birds in the state of Michigan'.[10] The bird was later adopted as the state bird for Connecticut in 1943 and Wisconsin in 1949, having previously emerged as the most popular bird in a survey of Wisconsin's schoolchildren in 1926 and 1927.[11] Cities too have been known to adopt an avian representative. In 2011, the Oriental magpie-robin was in with a chance of becoming the first 'Bird of Mumbai' when Mumbaikars were encouraged to vote from four short-listed birds. The campaign had a clear objective: to find a species that not only epitomized the city, but had managed to survive the city's rampant development. The organizer was quoted in the *Times of India* as saying:

> The idea is to elect a flagship species that showcases the never-say-die spirit of Mumbai and its surrounding areas . . . Despite pollution, thinning green cover and ever-increasing human population, several bird species have thrived on the alterations to their habitats. We want to elect one such species.[12]

Unfortunately for the Oriental magpie-robin, it was pipped to the post by the colourful coppersmith barbet, a tiny bird of brilliant green and crimson, known for its 'tuk tuk tuk' call.

Disagreements about what constitutes a national or flagship bird abound. Should it be selected based on popularity, or perhaps, like the coppersmith barbet, because of its especially unique or aesthetically pleasing appearance? Or should a national bird be one somehow befitting of character? The robin's triumph in the British polls certainly prompted some consideration of what a bird might say about the nation. Commentators were quick to remark on the robin's darker side – its 'alter ego', as David Lindo put it.[13] As Philip Hoare remarked in *The Guardian*, 'The robin is brutish, ruthless, and ready to ruck.' Could it be, he wondered, that 'over gentler contenders, we have plumped for the bird that we deserve?'[14] Meanwhile, Anna van Praagh of *The Telegraph* wondered whether the robin might be too ordinary to be the national bird of Britain.[15] So followed the questions: why not choose a bird that would benefit from the spotlight – perhaps one of Britain's red-listed species? Why not choose a more unique bird or one that Britain has played an especially important role in protecting? Of course, it is all a matter of opinion but the concern over the robin's somewhat 'brutish' behaviour raised some eyebrows. Could the gardener's friend and symbol of Christmas cheer really have such a dark side?

This was not the first time the robin's 'character' has been brought into question. The archives are full of 'horrified' accounts of the bird's apparent 'thirst for blood'. One such account appeared on the 18 June 1844, when it was announced in the *Belfast News Letter* that the robin redbreast was no gentleman. The published letter depicts a scene in 'charmingly wooded' grounds and a secluded spot, where 'an invalid lady used to delight to sit and read'. She had been overjoyed to find herself in the company of a robin, who was 'regular in his attendance upon her'. But the joy was short-lived, for the gardener identified the bird as 'an impudent little scoundrel' whose character had been blasted years before. Having nested in the greenhouse for a number of years, it was reported that one summer, on finding that his 'wife' refused to leave at the end of the breeding season, the robin promptly killed her – 'much to the horror of his admirers' – to regain his bachelor territory.

The robin's 'savage laws of divorce' were the focus of a piece in the *Evening Standard*, when in 1933 Eric Hardy outlined why naturalists such as himself wanted the robin removed from Christmas cards. As the case was described, not only was this a bird that failed to feature in any of the scriptures, but it was also a bird that will kill a rival, 'perch upon his victim's remains' and drive away his wife should she dare challenge his authority. As the article put it, 'Beneath the red breast of the robin lies the fiercest temper of all our garden birds.' Hardy and like-minded naturalists feared that the British nation – and the idea of Christmas – had been conquered by a bad-tempered bird.

The debates concerning the suitability of the robin as a national bird say a lot about anthropomorphism and the moralizing discourses that shape animal lives, but it also draws attention to another geographical concern: territory. All year round, robins protect their territory, which needs to provide a good and continuous supply of food and a place to rest or take cover. In the breeding season, this needs to include a good choice of nesting sites. A pair of robins will necessarily share a territory when breeding, but this partnership is short lived and in the winter males and females will part ways and hold their own territories. More often than not, the male will keep the territory that it occupied during the breeding season, and the female will leave, hence the less than savoury accounts of 'divorce' in instances where the female has been unwilling to give ground. Territory is a serious business, and as a frequently cited proverb from the third century goes, *Unicum arbustum haud alit duos Erithacos* ('A single bush does not shelter two robins').[16]

In his unfinished notes on 'Individuality in Birds', the ornithologist George W. Temperley described how his garden robin,

'One bush does not shelter two robins', illustrated frontispiece, woodcut from Roberto Valturio, *De re militari*, printed by Christian Wechel in Paris (1532).

Robert of Restharrow, had been replaced by another whom he called 'The Usurper'. Unlike Robert, The Usurper had darker feathers and was less inclined to eat from Temperley's hand. After several weeks of presuming Robert to be dead, he had whistled in the corner of the garden for The Usurper, only to see Robert land in the neighbour's garden. Robert was not dead after all but had instead been ousted by a rival who had taken his territory. Following an arrangement with the neighbour, Temperley had the fence rail removed so that he could cross into Robert's new territory and stand on the boundary with his arms outstretched. Robert perched on one hand and The Usurper on the other. As Temperley described it, the birds 'knew their own frontier to a hair's breadth and I could seldom get either of them to cross it'.[17]

The robin's song might have proven rich subject-matter for poetry, but it has a far more practical purpose: the defence of territory. Both male and female birds protect and hold territory and a good perch where a robin can be both seen and heard is an excellent vantage point from which to guard the borders and warn intruders off. If these signs are ignored, then a more aggressive response is required. At first, this might involve posturing – a puffed-out chest, with head tilted upwards and tail cocked. If this fails, a chase might ensue, and if the message is still not received, then a fight may be in order.

It is the ferocity of territorial disputes that has earned robins their somewhat violent reputation. This less palatable side came as a shock to the author of *The Secret Garden*, who recounted how her own charmed interactions with her little English robin were disrupted by an unwelcome intruder. Having gone out to her rose garden and begun her usual conversation with her beloved robin, she had been confused by his unusually shy demeanour and perturbed by his failure to reply in his usual conversational way. Burnett's confusion was cleared up when, in her own words, 'Out of nowhere darted a little scarlet flame of frenzy – Tweetie himself – with his feathers ruffled and on fire with fury. The robin on the branch actually was an Imposter and Tweetie had discovered him red-breasted if not red-handed with crime.' Witnessing Tweetie chase the intruder off with 'blood-thirst in his eye', Burnett declared herself quite pale with fright after beholding such 'righteous wrath'.[18]

When robins engage in combat they latch onto each other's feet, while striking blows to the head and body – an encounter that uses up a significant amount of energy and can prove lethal for one or both birds.[19] Such is the spectacle that in 1849 the Irish naturalist William Thompson transcribed several anecdotes in *The Natural History of Ireland*, each of which depicted the robin's so-called pugnacity. This includes an example from Merville in County Antrim, where a robin

> kept possession of the greenhouse, and killed every intruder of its own species (amounting to about two dozen) that entered it. This had been so frequently done, that an examination of two or three of the victims was made, to ascertain the cause of death; and a deep wound was found in the neck of each, evidently made by the bill of the slayer. The lady of the house hearing of the bird's cruelty, had the sharp point of its beak cut off; and no more of its brethren were slaughtered; but it did not long survive this slight mutilation.

In a second example:

> Robins being so wholly absorbed during combat as to be regardless of all else was ludicrously evinced at Springvale by a pair fighting from the air downwards to the earth, until they disappeared in a man's hat that happened to be lying on the ground, and in which they were both captured.

In a final anecdote, Thompson was taken with an account of two robins caught fighting in Belfast that had to be kept in cages all night to prevent deadly combat. While one of them was released the following morning, the tamer of the two was kept inside with the intention of permanently retaining it. When the retained bird appeared to be unhappy, it was released, only to be promptly rescued again after being subject to another attack. The attacker was driven off before a third release was attempted, but this was to prove fatal after the 'wicked and pertinacious antagonist' attacked and killed it after apparently lying in wait. It was noted that the 'tamer bird, though the inferior of the other in strength, always "joined issue", and fought to the best of its poor ability'.[20]

Perhaps the most humorous account of territorial dispute comes from Thompson's own observation. By his account, two fighting robins, having taken pause to recover breath, were separated by a duck that had happened to witness their 'wicked' combat. As Thompson described: 'In the most gentle and pacific manner [the duck] shoved with its bill the one to the right and the other to the left, thus evidently separating them to prevent a renewal of the conflict.'[21] Unfortunately, this was the end of the observation and no further word was offered as to whether the duck's efforts were successful in keeping the robins apart.

Such anecdotes reveal the significance of territory and why the relocation of a robin is unlikely to achieve a truce. Robins notice when other robins stop singing. In experiments where territorial males were trapped and removed from their territory, it was observed that their release prompted them to sing excessively in a bid to apparently reclaim and restore it.[22] When a robin disappears, less dominant robins, or juveniles that are looking to establish better territories, are quick to move in and claim the territory for themselves, while robins with neighbouring territories might also look to expand.

The robin's propensity for combat gave rise to David Lack's 'adventures with a stuffed robin'. For the bargain price of one shilling, Lack bought 'an exceedingly shop-soiled stuffed specimen' to play a starring role in his study of territorial behaviour.[23] He wired a stuffed robin to a branch in close proximity to pairs of nesting robins so that he could observe their response. Lack's studies found that individual robins attacked with different levels of ferocity, some only posturing and some attacking more

Two robins clashing while protecting their territory.

aggressively, depending on the time of year. The specimen met its end when a hen robin attacked with such force that she removed the specimen's head. This rather undignified end gave Lack an idea for further experiments, and the headless robin was gradually dismembered to see how much of the robin needed to be present to illicit a response. Surprisingly, he found that a bundle of red feathers was all that was needed to provoke a response in most territorial robins. To further assess the significance of colour, another robin specimen had its red breast covered. As suspected, it provoked little to no response from other robins.

The territorial behaviour of robins means that it is not uncommon for robins to attack their own reflections. Robins flying at car wing mirrors and tapping on glass have led many people to turn to the Internet to ask: why is a robin attacking my window?

In the middle of July, while taking a break from writing, I was standing at my office window watching a robin on the garden path, when I seemed to catch her attention. She flew at the window and for a brief second looked as though she was going to hit it at full speed. At the last moment she pulled up, hovered in front of my face and then flew away. The robin had flown at the window with intent, but it was too high for her to have seen her own reflection. On the spot she had flown at was a card from my friend's daughter and on the back a brightly coloured giraffe in orange felt-tip pen: a potential intruder. Given their connection to death, it is not hard to see why a robin's appearance at the window or tapping on glass might take on especial meaning.

In winter, robins are observed to be somewhat more tolerant of intruders. When food is scarce and energy is low, territorial spats are a highly risky business as the defence of perimeters requires a significant amount of effort. But this doesn't eliminate the chance of territorial dispute entirely. In 2011 the fierce territorial nature of the robin was the subject of Ross Hoddinott's

A robin sizes up its reflection.

Ross Hoddinott, *Territorial Strut*, 2011.

highly commended entry for the Wildlife Photographer of the Year competition. Hoddinott's aptly named *Territorial Strut* superbly captured an individual's warning pose – the cocking of its tail and aggressive posturing to better display its red breast – as it defended its territory in late December in a garden in Devon, England. Hoddinott had noted as many as eight robins in one patch – a high concentration for such a territorial bird.[24]

Robins in Britain are known for holding territories all year round and often move less than 5 kilometres (3 mi.) from their breeding territories, if at all, but robins across the European continent will migrate to warmer climes for the winter where food is more readily available. For this reason, Britain might see an influx of robins, especially along the east coast, while northern robin populations from Scandinavia, eastern Europe, Russia, Belarus, Moldova and northern Ukraine will move south or southwest to spend the winter in western, southern or central Europe and around the Mediterranean.[25]

In David Lack's study of ringed robins at Dartington, it was found that females were less present in the study area over the winter, which suggests that they were pushed out by more dominant birds. It has also been noted that the small numbers of British birds that do migrate are predominantly female, many of which have been recorded somewhere between the Netherlands and southern Spain. According to Peter Clement, the longest distance a robin has been known to travel is 2,400 kilometres (1,500 mi.) – from Belgium, where it was ringed, to the northern fringes of the Sahara Desert.[26]

The migratory comings and goings of robins were famously observed by Aristotle in ancient Greece, although he wasn't aware that this was what he was witnessing. Instead, the appearance of the robin in winter led him to develop his much-cited theory of transmutation. Having observed that redstarts disappeared from Greece at the end of the summer, just as robins appeared, he declared that one bird clearly morphed into the other:

> The rithacus (or redbreast) and the so-called redstart change into one another; the former is a winter bird, the latter a summer one, and the difference between them is practically limited to the coloration of their plumage . . .

Redstart (*Phoenicurus phoenicurus*), the bird Aristotle believed transformed into a robin.

> That these birds, two in name, are one in reality is proved by the fact that at the period when the change is in progress each one has been seen with the change as yet incomplete. It is not so very strange that in these cases there is a change in note and in plumage, for even the ring-dove ceases to coo in winter, and recommences cooing when spring comes in.[27]

Of course, much is now known about the phenomenon of bird migration, which is what Aristotle was actually witnessing. Indeed, it was thanks to a series of German experiments involving the robin that we know so much about how birds migrate.

Although there had been speculation that birds might use Earth's magnetic field as a means to navigate themselves, it wasn't until the 1960s that this was scientifically proven. It had long been believed that birds orientated themselves by using only the Sun and stars. However, following experiments that suggested that some night-migrating birds were able to navigate without access to celestial cues, Friedrich Wilhelm Merkel and his graduate student Wolfgang Wiltschko set out to replicate the experiments at Goethe-Universität in Frankfurt and find out what the birds were using instead. Robins were placed in a steel chamber: an orientation cage. This was initially done to disrupt their internal clocks and restrict access to cues, although the chamber also happened to partially shield them from Earth's magnetic field. The birds were disorientated to begin with but after several days had adjusted and were found to be perfectly orientated. As celestial cues were not available to the robins, they realized that the birds had become accustomed to the weaker magnetic field.[28]

To further test this theory, Merkel and Wiltschko performed further experiments with the orientation cage whereby they purposely altered the orientation of the magnetic field using a large electro-magnetic coil. The robins naturally hopped in a southwesterly direction, which was the direction in which they would usually head for their autumn migration from Germany. When the field was reversed, they altered their direction accordingly, demonstrating their ability to detect the altered magnetic field.[29] Merkel and Wiltschko published their important findings in 1965 and 1966.

Whether concerning matters of migration, territory or the imagined communities built around nations, states and cities, it would seem that the robin can tell us a lot about geography.

6 The Colour Red and a Christmas Story

And in autumn, when I sit and sing in the flaming leaves, little children laugh and clap their hands, for they hear the feet of Christmas coming.[1]

Red is central to the robin's story. As Michel Pastoureau suggests, the perception of colour is not just a matter of biology or neuro-biology; rather, it is a matter of culture: it stirs the imagination, rouses memories and inspires feelings that draw on a wealth of inherited knowledges and societal cues.[2] An examination of red's history reveals a colour of 'ambivalent symbolism' that occupies a prominent position within the cultural imagination. It is connected with life and death, love and violence, war and charity, and is especially resonant in Christian symbolism, where it is largely associated with blood and fire.

The symbolic significance of the colour red is evident in the origin stories of the robin's red breast. According to some accounts, it was through spilt blood that the robin gained its vibrant colour. One of the most prevalent of these tales concerns the blood of Christ. The robin, it is said, acquired its red breast when it came to the aid of Jesus on the cross and was splashed with a drop of Christ's blood. In some versions the robin attempted to remove the thorns from Jesus' head, while in others it had tried to comfort Christ with its song. While sacrificial reds are often transitory – either cleaned away or drunk – in these accounts the robin carries the sacrificial blood of Christ as a permanent marker of compassion.[3] Spilt blood is often associated with violence, but in the case of the blood of Christ it is a red that also sanctifies and gives life.

Close-up view of the Rosette Nebula. The red colour comes from hydrogen. Historically, the colour red has been considered a life force.

Ernest Hébert, *Virgin with a Robin*, *c.* 1880, oil painting. A robin appears behind the Virgin Mary. Associated with the story of the Crucifixion, robins were a reminder of what was to come.

The tale of the robin and Christ was the basis for a short story in *Christ Legends* by the Swedish author Selma Lagerlöf, the first woman to win the Nobel Prize in Literature in 1909. Lagerlöf tells the story of the world's creation and a little grey bird to which the Lord granted the name 'Robin Redbreast'. Having observed the colourful creatures around him – the parrots, the butterflies and the goldfish – Robin Redbreast asked the Lord why he was without colour and was told that his red breast feathers must be earned. For many years, and to no avail, it was said that generations of Robin Redbreasts tried all they could to earn the plumage that would befit their name. Then, one morning, on the outskirts of Jerusalem, a Robin Redbreast was shielding his family from a crowd of people that had gathered to crucify three criminals. Noticing that one of the three had been given a crown of thorns, the bird took pity on the suffering prisoner and, not being strong enough to remove the nails, removed the thorns from the man's forehead instead. As Robin Redbreast did so, a drop of blood fell on his breast and spread quickly across his breast feathers. In coming to the aid of Christ, the bird had finally earned his colour.[4]

A line from Lagerlöf's story, which underlined the robin's reward for his courage and compassion, prefaced Jo Nesbø's crime novel *Rødstrupe* (The Redbreast), which finds the detective Harry Hole on the trail of Norwegian neo-Nazis and Second World War Nazi sympathizers.[5] Alongside the book's reference to *Christ Legends*, *Rødstrupe* also mentions the brave and calculated risk that a minority of Norwegian robins take when they choose to stay rather than migrate for the winter – a decision that can be a matter of life or death.

In Britain, the robin's connection to Christ is commonly cited as one of a number of explanations for the robin's ubiquity at Christmas, but elsewhere the robin's story is more readily associated with Easter and features in Easter bulletins and sermons as a reminder of the importance of charity, compassion and faith. There are many variations. While the bird of Lagerlöf's story is desperate to live up to its 'redbreast' name, in other stories the spilt blood and the bird's colouration marks the birth of a new species; the point at which a brown bird gains a new name. In some accounts, the bird attempts to wash off the blood in a pool of water, only to find that the colour intensifies and spreads to become a vibrant, permanent mark of the bird's compassion. This rich symbolism was the focus of an Italian poem by Enrico Pea (1881–1958), 'The Robin' ('Il Pettirosso'), in which Pea described the robin as the bird 'who carries the insignia/ of Christ on his immaculate chest'.

In the United States, the Easter robin also features in Christian stories for children, where it is the American robin that is depicted in accompanying illustrations. In one example, the story of the Easter robin is described as a Pennsylvanian Dutch tale, thus linking the story back to the bird's European 'cousin'.[6] Although stories of the robin and bloodshed are likely pagan in origin, the robin's connection with the Easter story is often wrongly assumed to originate with the Bible. In this matter, the case of the Easter robin is perhaps especially confounding in North America, given that the story of the crucifixion is a considerable way out of the American robin's range. As is the case with many tales and legends, the origins of such stories are frequently lost, having been passed down through generations and adapted along the way.

Tales of bloodshed are found elsewhere. For instance, a story from the Nyoongar people in southwestern Australia tells of a fight over hunting rights between a willy wagtail – or so-named 'chitty chitty' – and a scarlet robin. In the Dreaming, it is said that the scarlet robin bled from the beak after being hit by the willy wagtail, which forever reddened its breast.[7] To this day it is alleged that the willy wagtail has a monopoly over the best hunting grounds and can be frequently seen chasing away the scarlet robin.

Bloodshed might be common to robin origin stories, but there are other tales that suggest an alternative: that the robin got too close to a fire. Here is another of red's close associations and yet

Vincenzo Leonardi, European robin, *Erithacus rubecula* (L. 1758), from *A 17th-Century Handbook of Bird-Care and Folklore*.

another that is resonant to Christianity. A robin was said to have kept the baby Jesus warm while he lay in the manger. Desperate for help and struggling with the cold, Mary pleaded with the stable animals, but they could not or would not help. As the last embers of the fire were dying away, a robin came to Mary's aid and fanned the flames with its wings. It was when adding new sticks to the fire that the robin was said to have moved too close to the flames and singed its breast feathers. Once again, the robin was to be forever marked by its compassion and charity.

Similar stories are found in Native American mythology. In a Miwok legend, Wit-tah-bah the Robin gained his redbreast when he spread himself over the embers of a fire to protect it from theft.[8] In the Mi'kmaq legend about fire, a robin was asked to fly to a place where lightning had hit the ground in order to bring sparks back to make a fire for a wise woman, Nukumi. The sparks were too hot for the robin to carry so he used two dry sticks to hold the flames for his flight home. As the robin flew back, the sticks burnt and the robin's breast turned red. He brought the fire to the wise woman and added additional wood to create a Great Spirit Fire. It is said that from that day forth all robins had red breasts.[9]

Much like the robin's breast feathers, fire is rarely red. Yet in standard descriptions of the colour, red is regularly described as being the colour of fire. This, as Michel Pastoureau suggests, is perhaps owing to the idea in ancient societies that red was the colour of life and fire was a living being. Because fire was considered a source of light and life, it was deemed a bad omen if it was extinguished. Indeed, the colour red's enduring association with life and the warmth of fire can be seen in another robin tale, which doesn't involve the small bird getting burnt. In his talk to the Anthropological Society of Washington in 1888, Robert Fletcher drew attention to a poem by George Gascoigne – 'The Complaint of Philomene' (1576) – as evidence of a curious legend. It describes how merlins, a small species of falcon, capture robins during the winter and hold them prisoners so that the merlins might be kept warm by the robins' glowing red breasts during the cold nights.[10]

Fire's symbolism is not straightforwardly good. Within Christianity it is also associated with hell, where fire is destructive and cruel. In the *History of European Morals* (1869), the Irish historian William Edward Hartpole Lecky cites one popular legend in which the redbreast 'was commissioned by the Deity to carry a drop of water to the souls of unbaptised infants in hell'. It was during this consignment to hell that the robin was said to have singed its breast in putting out the flames, taking on its fiery colours as a result. Lecky cites this legend in a passage in which he notes the tendency of Catholic priests to preoccupy themselves with hideous beings, 'ghastly pictures of future misery' and a 'yawning hell' that was ever ready to receive its victims. He notes the link to the crucifixion of Christ in a footnote.[11]

It is perhaps because of red's strong associations with fire and blood, and their symbolic significance to Christianity, that the robin has featured in so many Christian stories. For instance, the robin makes an appearance in C. S. Lewis's *The Lion, the Witch and the Wardrobe*, a children's story known for its strong Christian message. In a crucial scene, a robin appears just as the Pevensie siblings seem to be lost in the forest and unsure of their direction. The bird, which couldn't have had a 'redder chest or a brighter eye', not only guides the children through the forest, but leads them to an important contact. As one of the boys wonders aloud whether the robin can be trusted, his brother reasons: 'They're good birds in all the stories I've ever read. I'm sure a robin wouldn't be on the wrong side.'[12]

If the tales of fire and spilt blood in explanations of the robin's colouration aren't to your satisfaction, then you may wish to know that, according to scientific explanation, a robin's colour comes from two types of melanin pigment: the darker tones from eumelanin and the red and orange tones from phaeomelanin. Albino robins, which have no melanin at all, have grey upper parts, while their much celebrated 'red' breast is almost completely white.[13] An albino robin sighting is rare; they are thought to have low survival rates, for they are easy for predators to spot and their eyes are sensitive to light. There is also the matter of their red feathers being vital if they are to effectively signal to other robins in territorial disputes.

Whether as a result of blood, fire or melanin, such is the robin's strong association with red that it has lent its name to all manner of 'things'. Take, for example, the glossy-red photinia or 'red robin' hedge plant, the Robin Redbreast procumbent rose or the ragged robin wildflower, which, while pink, is named as such because of the vibrancy of its colour. The robin has also lent its name to red wines, bars, a marine sponge and even a fictional

An albino robin with breast feathers that are almost white.

superhero. While there has been some debate about the origin story of Batman's sidekick, *The Untold Legend of the Batman*, which was published in 1980, claimed that the American vigilante Robin gained his name because of the colour of his costume, which was observed to be 'as brilliant as a robin redbreast'. When Dick Grayson took up the role as Batman's sidekick he considered the 'Robin' name to be perfect; like Batman, he too would have the name of a flying creature, but he would also share the name with one of his heroes, Robin Hood. (Beyond demonstrating the popularity of the name, there is no connection between Robin Hood and the bird.)

Because red is regularly associated with warriors, martyrdom and aggression, it is also a popular colour with sports teams. Five teams in the English Football League – Bristol City, Crewe Alexandra, Cheltenham Town, Wrexham and Swindon Town – as well as the rugby team Hull Kingston Rovers, are (or have been) affectionately known as 'the Robins' on account of their red strips. Swindon Town changed their shirts to their more familiar red in 1901. While initially a dark maroon, a lighter shade was chosen for the start of the 1902–3 season. This resulted in the club's nickname 'the Robins' appearing in print for the first

The *Photinia* hybrid 'Red Robin', *Photinia* x *Fraseri*.

Carnaby Street robin-themed Christmas lights, London, 2013.

time in programme notes.[14] The team's mascot is the larger-than-life Rockin' Robin, which was also the name of Wrexham AFC's mascot until he was given the 'red card' in 2001. While Rockin' Robin's 'distinctive buoyant shape and chirpy demeanour' were said to be loved by many, he was replaced by Wrex the Dragon and a new nickname was adopted. A switch to red shirts also gained Cheltenham Town their 'robins' nickname, which set the stage for club mascot Whaddney the Robin, while it is Scrumpy the Robin that has the job of getting the crowds going for Bristol City FC. Curiously, while Crewe Alexandra is also known for its 'robins' nickname, it is Gresty the Lion who has been entrusted with the important task of rallying the crowds.

Beyond superheroes and sports teams, perhaps the most important connection with red uniforms concerns the attire worn by Victorian letter carriers, for it is this association that has secured the robin's status as a Christmas icon. If visiting the UK for Christmas you will find the country engulfed by robins – and not just the kind you find in the garden. Robins hang from Christmas trees, burst out of Christmas crackers, market all kinds of festive confectionary and merchandise, and are splashed across the postal network. If these robins aren't enough, you could get a robin tea towel, a pair of robin earrings or perhaps even some robin tableware for Christmas Day. Since commencing this book, I have been spoilt with all manner of robin gifts – lights, cards, chocolate, socks, tree decorations – and nearly all of them have been Christmas-themed.

There are a number of reasons for the robin takeover and its long association with Christmas, including its Christian origin stories. But while the robin pops up in many a festive scene across its ranges, its utter ubiquity at Christmas is an especially British affair. On noticing my choice of Christmas jumper, which was suitably covered in robins, an American colleague once confessed to having 'wondered what the deal was – why so many robins?'

Red tailcoats with blue lapels and beaver hats were worn by letter carriers in London as early as 1793, when they were unveiled to mark the birthday of Queen Charlotte, wife of King George III. By 1834 the uniform was rolled out to other large towns and cities. Because of its associations with power, red is widely considered a regal colour and was thus fitting for postal workers of the Royal Mail. However, the uniforms did far more than give letter carriers a majestic appearance. The scarlet uniforms also gained them an affectionate nickname: 'robins'. This name is in evidence in Anthony Trollope's novel *Framley Parsonage*, which includes the character of Robin postman who, on one occasion, is invited into the parsonage for tea and buttery toast in a bid to persuade him to take a letter elsewhere.[15] (Trollope himself had worked for the London Post Office as a clerk from 1834 until 1841, a position that he was said to have hated).[16] As further evidence of the popularity of this affectionate name, Alison Uttley's magical series of children's books, which starred Little Grey Rabbit and her woodland

A letter carrier in uniform, late 19th century, hand-coloured non-photographic lantern slide.

friends, also featured the character of Robin postman – an actual robin redbreast – who would 'tap, tap, tap' on the door to deliver the animals' letters.[17]

The postal service was largely used by the wealthy because of the cost of receiving post (the recipient paid for the postage on

UK Christmas stamp, 2016.

Isle of Man Christmas stamp, 1995.

delivery), but it was reformed in 1840 with the introduction of the self-adhesive penny black stamp – the world's first postage stamp. This meant that any letter weighing less than half an ounce (around 15 grams) could be sent anywhere in the country for just one penny. In 1840 alone, 169 million pieces of post were delivered, which was more than double the number sent in the previous year. By the turn of the century, this was 2.74 billion.[18] The introduction of the penny black stamp revolutionized communication and sending messages by post became increasingly popular. It also paved the way for the robin's emergence as a ubiquitous Christmas icon. Three years after the introduction of the stamp, the first commercially printed Christmas card was sent. Designed by J. C. Horsley at the request of the civil servant Henry Cole, the Christmas card was designed primarily as a sentimental and social object that was intended to further encourage people to connect with each other through the post. The circulation of Christmas cards revealed ties and networks of social obligation, but they also offered good cheer, happy memories and well wishes from loved ones such that their arrival would be keenly anticipated.[19]

It is easy to see how the red-breasted bird made its way into a familiar range of Christmas iconography, for it was the letter carriers, or 'robins', who brought Christmas greetings to people's doors. Postal workers were recognized as 'essential intermediaries in the process of creating and maintaining a merry Christmas'.[20] Letter boxes weren't introduced to houses until 1852, so each letter was handed directly to the intended recipient as part of a highly personalized service. Early cards depicted robins carrying the Christmas post in their beaks or perched atop letter boxes, in some cases even donning the postal uniform. Across the nineteenth century, cards with pictures of postal workers and postboxes were particularly fashionable. In fact, while it was the development of the Christmas card that was to bring the robin long-lasting commercial success, the bird had originally appeared on valentine cards and other stationery for more general use, including letterheads, envelopes and notelets.

As Patricia Zakreski, a scholar in Victorian culture, suggests, the Christmas card has a complicated history as an object of both social and commercial importance. Even while the first Christmas card was developed as a sentimental object, it already revealed an awareness of industrial modernity and a growing consumer culture. A dotted line above and below the picture on Horsley's first Christmas card required that the sender do no more than enter the names: To/From. Time was precious and the Christmas card offered an easy and efficient way to send greetings en masse. In the second half of the nineteenth century, new forms of mechanization made mass colour printing of different designs cheaper and easier and so the Christmas card's commercial significance was confirmed, along with a range of Christmas icons, including the robin.

As the Christmas card became more ubiquitous, a focus on aesthetics and design became more and more important to its continued success. Robins in wintry scenes, robin carollers and

Christmas card to Henry Cole, sent by John Callcott Horsley, 'Xmasse 1843'. Hand-coloured lithograph illustrating a feast scene and bearing an image of an artist, by way of signature.

Poster advising on the benefits of posting early over the festive season, featuring a robin carrying a letter.

hearth-side robins might be expected, but some of the designs might raise an eyebrow. Perhaps a standout example is a set of cards from the 1880s that included a beautiful illustration of a dead robin, accompanied with cheerful messages of merriment and joy. There are a variety of different ideas about the meaning behind such seemingly macabre images. Some suggest that dead birds were a symbol of good luck, an idea that is often attributed to the killing of wrens on St Stephen's Day, or 'wren day', in Ireland, although there are various traditions across Europe with different origin stories. (In some traditions, the wren is seen as a symbol of the passing year, while in others it is linked to the Christian story of the King of All Birds, where a treacherous – some might say clever – wren cheated to win the title.[21]) If not good luck, then it has also been noted that death was a prominent feature of Victorian life, as were posthumous portraits, with dead robins perhaps representing the fragility of life and their childlike innocence drawing parallels with *The Babes in the Wood*. As histories of the Christmas card have suggested, the morbid picture of a beloved robin was bound to elicit sympathy and was most likely a reference to stories of impoverished children freezing to death at Christmas.[22] Indeed, by the end of the nineteenth century, destitute children were frequently referred to as 'poor robins'.[23]

References to morality and social reform were common to the Victorian Christmas, as was charity. While festive charity has been in evidence since the medieval period, in late Victorian England, charitable work was specifically concerned with the provision of a festive experience for the poor.[24] In this context, a Christmas card with a robin on it played an important role in the establishment of a movement that aimed to provide free Christmas dinners, known as 'robin dinners', for destitute children. In recounting the story of the movement, Reverend Charles Bullock of London offered the following:

> About twenty years ago a Christmas Card – a special Christmas Card it proved to be – came into the writer's hands. It had a picture of a Robin Redbreast at the top,

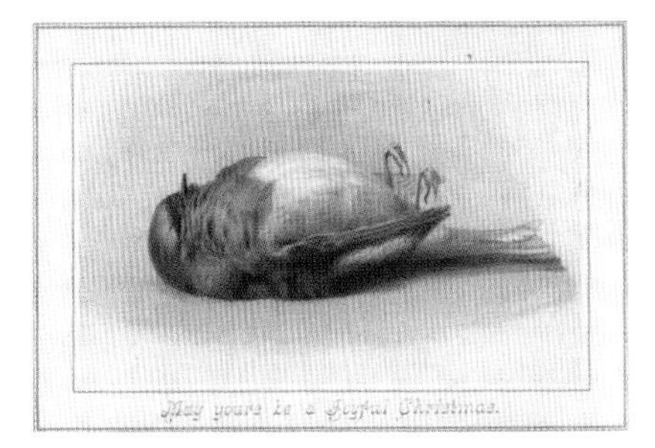

Dead birds on Christmas cards were not so unusual in the 1880s.

'A Merry Christmas and a Happy New Year'. Robins were a common feature of Victorian Christmas cards.

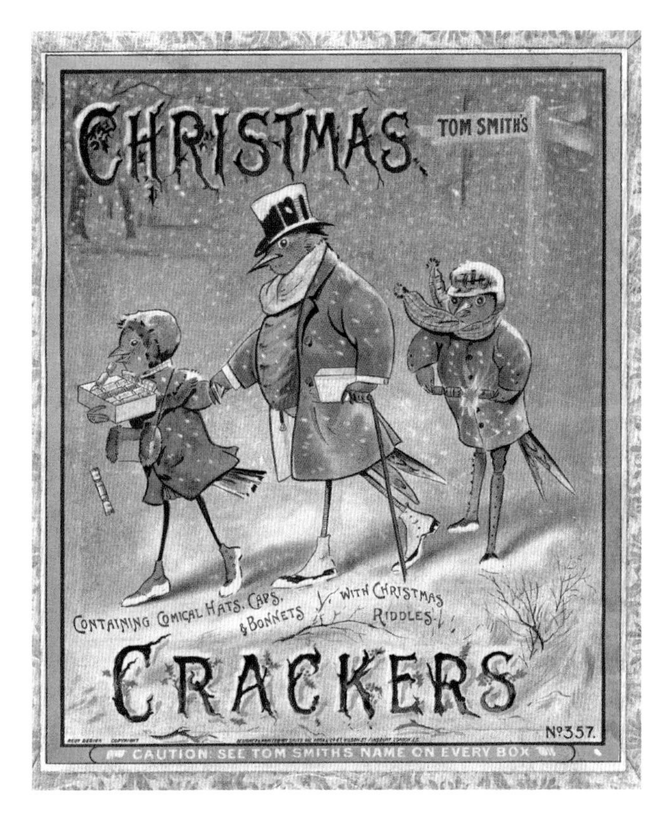

Victorian Christmas crackers made by Tom Smith & Co, *c.* 1870–90.

presenting himself on a snowy Christmas morning at the window-sill, with a letter in his beak bearing the message 'A Merry Christmas to you.' The picture seemed full of the poetry of love. The poor little bird was the needy one, but he came to give rather than receive.[25]

It was this Christmas card that inspired the Reverend to write 'Robin's Christmas Salutation', which contained the following lines from the robin:

You would give *me* some crumbs from your table I know,
And guard *me* from cold when the stormy winds blow;
Deal your bread to the hungry, and then your rich fare
Will be sweeter than ever, for *God* will be near.

In 1876 the Reverend published the hymn in the magazine *Hand and Heart*, requesting that readers take note of the robin's sermon and send monetary donations towards a Christmas meal for impoverished children. This resulted in the first 'robin dinner' and the establishment of the Robin Dinner movement. In its first year, the movement provided for between three hundred and four hundred children. Just one year later it served about 10,000. It was the Reverend's wish that the movement would extend far beyond London and become a national institution.[26] Across the next century, 'robin dinners' were a regular fixture of the festive season.

The winter robin's search for food has been the basis for numerous lessons in faith and charity. In 1855 Mrs Gatty, a keen naturalist and children's writer, published *Parables from Nature*.

'Robin breakfast' ticket of December 1910 in *The Guild of the Good Shepherd Robin Breakfast Book*.

Inspired by Hans Christian Andersen's fairy tales, but regretful of his tendency to leave so many of his stories 'without an object or moral', Mrs Gatty was convinced of the morals that could be learned from nature. The robin proved a worthy subject in the parable 'Daily Bread', which Gatty based on the words of Jesus in the sermon on the mount ('Your heavenly Father knoweth that ye have need of all these things', Matthew 6:32). In the bleak of winter, with a flash of red and a joyful song, a robin's 'cheerful composure' is contrasted with the gloomy disposition of a cast of other creatures who are far less optimistic about what the winter may hold. Confident that 'kind chance' would provide for him throughout the winter, the robin is rewarded with berries

H. Weir, *Robin on a Branch of Holly*, 1858, colour lithograph.

from the holly bush and crumbs and pieces of bacon fat that are scattered by a man cheered by his song. The lesson is clear: have faith and give thanks for each day's blessing and you will be provided for.[27]

The depiction of the robin as a figure of morality and Christian virtue is seen elsewhere in Victorian children's literature. Sarah Trimmer's *History of the Robins* (originally titled *Fabulous Histories*, 1786) was, in her own words, 'written with the benevolent design of teaching the young to be tender-hearted and compassionate, not only to the feathered race, but to all the animal creation'.[28] The story concerns a human family of four, the Bensons, and a pair of redbreasts that nest in their orchard and raise a family – Robin, Dicky, Flapsy and Pecksy. Across the course of the story, the young robins and the Benson children are taught about the importance of family as well as 'proper behaviour', both separately and through their interactions with one another as the birds grow, learn to fly and feed for themselves.

The children are taught to be charitable towards the robins, while the young robins are instructed to share the offerings from the children and grant them their song in return. Robin, the eldest of the young birds (and the bird most 'ruined' by his vanity), ignores his parents' instructions on learning to fly, is badly injured as a result and is eventually taken by the children to be looked after indoors, where he remains for the rest of his days. When the time comes to leave the nest for good, Dicky and Flapsy are keen to get away from their parents. They leave the orchard they called home, only to find themselves forever confined to an aviary after being captured in a trap-cage. Pecksy, however, who is happy with what she has, remains in the orchard and enjoys a very happy life. Of course, while Pecksy's decision to remain and be happy with her lot was celebrated in the story, in reality she would have likely been chased off in a territorial dispute.

While emphasizing Christian charity, Trimmer was clear that her story was not intended to encourage children to care for birds above humans and thus placed emphasis on instilling children with a sense of their place in the world – at the head of creation but behind their parents. So full of life were Trimmer's robins that she was firm in her introduction to the book that it did not contain the 'real conversations of birds' but merely a set of fables that should act as guidance.[29] It might come as no surprise that Trimmer's story was criticized for its distinction between the deserving and undeserving, something which could also be seen in her involvement in the Sunday School Movement, where she argued that Sunday schools for the poor offered an important means of instituting social control and thus reinforcing God's divine social hierarchy.[30] In an account of Beatrix Potter's early exposures and childhood reading, Potter was said to have inherited a copy of Trimmer's *History of the Robins* from her grandmother as an early primer. She was later said to have hated its moralism but unsurprisingly had no objection to its anthropomorphism.[31]

To return to the bird of winter, it is worth asking why robins are celebrated for their Christmas fortitude. An autumn robin might appear wistful and melancholy, but in the depths of winter, the robin, with its bright breast and song, has long brought warmth and cheer. Again, the significance of the robin's red breast shines through and seems to take on an especial intensity. As historian Michel Pastoureau suggests, it is not a coincidence that in both French and German the word for red (*rouge* and *rot* respectively) has been historically used as an adverb to express intensity in the same way as 'very'. For example, in the French language of the sixteenth and seventeenth centuries, *cet homme est rouge grand* meant 'That man is very tall.'[32]

In a nod towards the robin's ubiquity at Christmas, the 1906 December edition of the *English Illustrated Magazine* published

New Year postcard, December 1928.

a fictional dialogue between Robin Redbreast and his 'northerly cousin', Robin Bluebreast, in which the two express their exasperation over humankind's sentimentality. Redbreast mocks the idea of the gardener's friend: 'Ha! Ha! Ha! And the very gardener thinks I sit on his spade for the pleasure of his company!', while Bluebreast describes the love for Redbreast's song as nothing more than 'sentimental nonsense', for he knows that Redbreast's song is truly a war song and nothing to do with broken hearts or Christmas cheer. But why, Bluebreast asks, are Redbreast and Father Christmas always together? Redbreast informs him that this is no legend, for when the seasons change and the snow comes, redbreasts will appear in gardens in greater numbers and sometimes even homes; their red waistcoats even match the holly

The robin's 'northern cousin', the bluebreast (blue throat, *Luscinia svecica*).

that adorns the mantelpieces. And so, there can be no doubt that the seasonal presence of redbreasts is a reminder that Christmas is truly here.

The northerly cousin of Robin Redbreast was in fact a bluethroat (*Luscinia svecica*), flying from the northern regions of Scandinavia to winter in North Africa. It has an ultramarine blue throat, forehead and breast, which is edged with narrow bands of black and white and a band of rufous-coloured feathers that can have a burnt orange appearance. While it belongs to the genus *Luscinia*, it is another bird that once belonged to *Erithacus* along with the European robin, hence its description as a northerly cousin.[33] During their conversation, Bluebreast reflects on his own southerly migration for the winter and supposes that it is probably for the best that he is not responsible for bringing Christmas cheer, not least because there is no such thing as a sky-blue berry.[34] It would certainly seem that Robin Redbreast is the perfect icon for Christmas. Not only does he bring cheer and match the holly, he also complements Father Christmas – red is surely the colour of the season!

As the robin's Christmas iconicity would suggest, Robin Redbreast has benefited from many of red's symbolisms and inspired feelings. But in turn, the robin has shaped and motivated a set of quite disparate social and cultural imaginations to leave its mark on religion, consumer culture, morality tales and legends.

7 Robin Futures

According to multiple conservation indices, the European robin is categorized as a 'species of least concern'. In fact, robins seem to be increasing in numbers, as are American robins and clay-coloured thrushes. But this is not the case for all robins, and in an era of climate change, environmental degradation and rapid urbanization, nothing can be taken for granted.

According to the IUCN's Red List of Threatened Species, which offers comprehensive information on the global extinction risk status of animals, fungi and plants, the rufous-headed robin and Seychelles magpie-robin are both endangered, while the Swynnerton's robin is considered vulnerable. In 1960 it was thought that there were only twelve Seychelles magpie-robins left, which were restricted to Fregate Island in the Seychelles. Once thought to be common throughout the archipelago, the birds suffered catastrophic decline as a result of habitat destruction – most notably the clearance of nearly all natural forest to make way for agriculture on some of the islands – and the introduction of alien predators, with the overcollection of specimens noted as a potential contributory factor on some of the smaller islands. In 1990 a recovery programme was launched by BirdLife International and funded by the Royal Society for the Protection of Birds (RSPB). A combination of predator eradication, supplementary feeding, habitat restoration and a ban on some pesticides led to rapid

A robin sings at dusk.

increases in numbers, which enabled the translocation of breeding pairs to other islands. In 2005 the Seychelles magpie-robin was downgraded from critically endangered to endangered. In 2015 there were an estimated 280 individuals across five islands, all with tight biosecurity protocols.[1]

Perhaps even more remarkable is the story – or 'resurrection', as it was dubbed – of the Chatham Island black robin (*Petroica traversi*), which was down to a single breeding pair, the smallest possible viable number for species survival. The last remaining female, nicknamed 'Old Blue', became something of a New Zealand hero for her 'efforts' in securing the survival of her species, when a pioneering conservation approach used parents from other species to incubate her eggs (her first clutch was removed, and she would lay a second). As the ancestor of all living black robins, she had a plaque erected to commemorate her as the 'saviour' of her species. Old Blue's story was narrated as a moral fable for children in Mary Taylor's *Old Blue: The Rarest Bird in the World*, which serves as a story of hope, hard work and miracles. As animal scholars Philip Armstrong and Annie Potts suggest, the story of Old Blue has almost become a 'secular myth of redemption'.[2] Because of the nature of the success, black robins not only are celebrated but have become a focus for studies looking to understand the effects of inbreeding in conservation.

Elsewhere, for the European robin it would seem that little has changed. It has been a garden favourite for centuries, and in much of its range can still rely on an abundance of food, especially where garden feeders are available. But while the robin is as loved as always, and its status has changed very little, the robin's environment has altered substantially.

In urban environments, night-time singing of diurnal birds is not new, but the rise of such activity has been attributed to an array of possible factors, including light pollution. As urban areas

Seychelles magpie-robin (*Copsychus sechellarum*).

Chatham Island black robin (*Petroica traversi*).

continue to expand, and urban night-time economies become an important part of 24-hour societies, light pollution is increasing globally. Artificial lighting has been linked to changes in the daily timing of robin song, with robins singing earlier in the morning and later at night.[3] These changes have also been observed with the American robin, which has been historically cherished for its dawn chorus.[4] Following these observations, researchers have begun to investigate whether artificial lighting might have an impact on natural seasonal rhythms. In studies carried out in southern Germany, it was found that male robins were more likely to sing earlier in the season at sites affected by light pollution. A network of brain nuclei known as the song control system is responsible for song production and learning, and the regulation of the song control system is influenced by hormones such as melatonin, which facilitates sleep. Researchers in Germany from the Max Planck Institute for Ornithology suggested that artificial light reduces melatonin levels and so winter days that are lengthened with artificial lighting might be perceived as spring days, triggering earlier singing.

Much more research is needed to confirm the causal links of these observed effects, the consequences of which are little understood. Could singing earlier in the year lead to longer breeding seasons and an opportunity for additional robin broods? Might there be even more robins competing for territory? Or might singing over a longer period of time lead to increased stress levels or exhaustion, making predation a greater risk? If light pollution has an impact on survival rates, it is still unclear how.[5]

It has been found that daytime noise is also a strong predictor of nocturnal singing in urban robins. As urban areas continue to grow, so too has the emission of anthropogenic noise. Urban noise now occurs for longer periods of the day, comes from more powerful sources and occurs across a wider geography. This can have a dramatic impact on the transmission of animal acoustic communication and patterns of behaviour. A team of scientists from the University of Sheffield in the UK found that daytime noise levels were higher at locations where robins sang during the night. This suggested that robins were either singing into the night to reduce the time that they spent singing in competition with other acoustic sounds or taking advantage of the relative quiet of the night to give *additional* signalling. The research indicated that while light pollution might lead to a passive physiological response (the reduction of melatonin levels), nocturnal singing as a response to daytime noise was a behavioural adaptation.

If robins alter the timing of their singing in response to urban noise levels, then it is also thought that they might change the frequency and amplitude of their song according to studies done with other songbirds. Urban noise, such as that produced by traffic, tends to be low-frequency, which is particularly significant for robins. In 2015 researchers at Newcastle University in the UK published findings from a study on the impact of wind turbines

Mural by Yash, Stockholm, Sweden.

on songbird communication, offering an important reminder that noise pollution is not confined to urban areas. Noting that robins typically use low-frequency sounds in their song as a means of protecting their territory from intruders during hostile encounters, the study suggested that robins might be less able to deter rivals when their song is drowned out by anthropogenic noise.[6] It is thought that low-frequency sounds produced by robins make them sound bigger and thus reduces the need for physical encounters. In making it difficult for robins to be heard, the low-frequency noise from the wind turbines potentially increases the need for physical contact, which, as we know, can be highly risky and incredibly costly in terms of energy.

The Newcastle study is one of many to have examined the environmental impacts of wind turbines at a time when the need for more renewable energy is urgent. On this point, the study reveals the effect of wind turbines on bird life, while underlining the role that turbines can play in tackling climate change, which is itself a significant threat to biodiversity. With this in mind, the findings from the robin study are a reminder that the impact on birds goes beyond the risks posed by direct collision with the turbines and concludes that the measurement of noise pollution should be incorporated into environmental assessments to reduce the negative effects on wildlife.[7]

If acoustic and visual landscapes have changed beyond recognition, then so too have environmental protections. In 1979 the European Union Birds Directive and the Bern Convention made non-selective methods of bird trapping illegal. The Bern Convention is a binding international legal instrument, which was signed with the aim of promoting national conservation policies and protecting the natural heritage of the European continent and some parts of northern Africa.[8] While this was to have significant ramifications for robins and other migratory birds, the illegal

Ambelopoulia, a traditional Cypriot dish of cooked songbirds.

trapping of birds is still a big issue across the Mediterranean more than forty years on.

In Cyprus, *ambelopoulia* is a traditional dish that originally consisted of blackcaps (*Sylvia atricapilla*) grilled and enjoyed whole, but has been expanded to include about 22 other species of songbird, including the robin. This dish is still served in some restaurants, but is illegal, as it involves the trapping and killing of wild birds. The issue of illegal bird trapping is found across the Mediterranean, but Cyprus is regularly highlighted on account of the numbers caught because it is a key migratory route for many birds. As science journalist Shaoni Bhattacharya described it, it is an issue that encompasses crime, culture, politics and science; it is highly emotive and increasingly lucrative (a platter of a dozen birds can sell for 40–80 euros).[9]

The eating of *ambelopoulia* is a tradition that goes back centuries. The traditional trapping method involves the use of 'limesticks' – twigs that are coated in a sticky glue-like substance (bird lime), made by boiling the fruit of the Syrian plum tree. The twigs are then placed in bushes and when birds perch on them

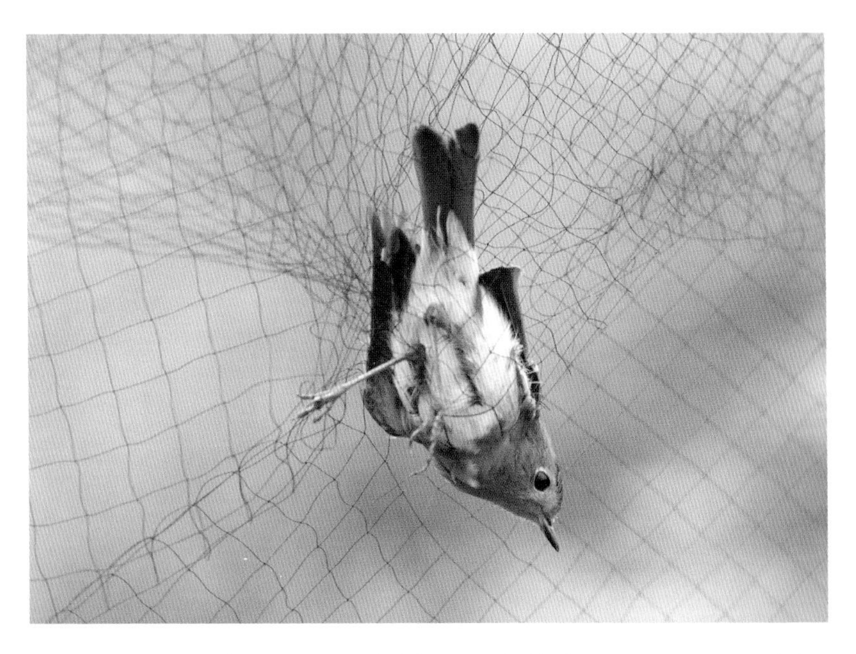

A robin is caught in a mist net.

they become stuck. The more the birds move, the more stuck they become. In 2013 a survey of bird trappers found that limesticks were generally used on a relatively small scale, for personal use or as an additional source of income. Mist nets, however, can trap birds on a much larger scale. This method involves cutting large tracts through native scrubland, planting trees likely to attract birds and then creating corridors across which nets can be strung. The nets are fine enough to go undetected, and birds become entangled when they fly into them. In many cases, taped birdsong is used as a lure to draw greater numbers.[10]

Bird lime features in a rather macabre folk tale from northern Portugal in which Sigli, a little boy who was known for his cruelty to birds, paid for his crimes with his life. Sigli, it was said, took great pleasure in catching, tormenting and killing all manner of creatures and took a particular pleasure in catching robins using bird lime. The extent of his killing was so great that King Robin of Birdland issued invitations to his 'feathered subjects' and 'beasts of the field' to a meeting in which they could discuss a means of punishment. A fox by the name of Reynard proposed that Sigli be killed using the same method that the boy had used to kill others, an idea that was especially supported by Mrs Queen Bee, whose hive had been destroyed by the boy and his father.

A figure of a man was made of bird lime and placed opposite the castle in which Sigli lived. On seeing it from the window, Sigli mistook the figure for a beggar. In a flurry of fury he ran out to

Unknown artist, *Still-Life with Wine Glass, Ladyfingers and Robin*, 1863.

the figure and kicked it, at once becoming stuck. The more he struggled the more firmly he was glued into place. His cries for help were heard by his father who vowed to squeeze the man to death for hurting his son and so too became stuck to the bird-lime figure. Having bound themselves so tightly to the frame, the two quickly died of suffocation.[11]

In 2017 the Bern Convention launched a social media campaign starring the European robin to promote public awareness of illegal bird killing. The campaign #thelasttweet followed four fictional birds, Sylvia Blackcap, Albert Goldfinch, Ninja Dove and Red the Robin, to tell the story of bird migration across Europe. The story of Red took followers on a journey from Finland to the warmer climes of Italy. Red made his way through Estonia, Latvia, Lithuania and Poland, alerted his followers to the latest advice on what to provide garden birds with during the winter and directed them to recent research on why songbirds are increasingly singing

Luis Meléndez (1716–1780), *Still-Life with Birds, Grapes, Pear, Apple, Potato, Figs and Nuts*, oil on canvas.

into the night. Red continued through the Czech Republic and Austria with the aim of reaching Italy to make his way through Venice, Verona and Bergamo. Here he was to meet up with his robin friends Bob and Rob for the winter. In his penultimate tweet Red informed his followers that he had found Bob caught in a trap in northern Italy – the tweet was quickly followed by the announcement of his own untimely death: his final tweet. In recognizing the great store that societies place in last wishes, the creators of the campaign decided that each bird would use a final tweet to make a plea for change. It was hoped that the message about their illegal killing and capture might take on greater poignancy as a consequence.[12]

Two years later, the Union of Czech and Slovak Zoological Gardens (UCSZOO) ran their own campaign against the illegal killing of migratory songbirds in the Mediterranean. Instead of addressing a domestic audience, the campaign targeted visitors from elsewhere in Europe, with large billboards placed along the key driving routes into the capital cities of Prague and Bratislava from their respective airports. While the killing of migratory songbirds affects hundreds of different species, like #thelasttweet, the UCSZOO campaign featured a robin and a goldfinch, two charismatic species with wide public appeal. (When the illegal killing of birds in the Mediterranean made headlines in the UK, it was similarly the threat posed to the robin – the nation's favourite bird – that was at the centre of the story.[13])

What is clear from a number of these campaigns is that, despite its abundant status, the robin has become something of a mascot for conservation issues on the grounds of its charismatic appeal – its charm, aesthetic qualities and cultural associations.[14] For instance, in the UK, bird charity SongBird Survival named 21 December National Robin Day in a bid to raise awareness of the declining numbers of songbirds and other small birds across the country. The robin was selected for the campaign because of its national popularity and familiar presence in winter gardens.[15]

Another example of the robin's enduring popularity can be seen in the debate over the use of UK government-issued licences to cull wild birds, otherwise protected under the Wildlife and Countryside Act 1981. This became a matter of public concern when, in 2018, environmental campaigner Jason Enfield submitted a freedom of information request to ascertain how many licences had been issued by Natural England to permit the killing of wild birds. While the request had initially been submitted as part of a campaign to halt the cull of English ravens, it revealed that at least forty species of birds were the focus of individual licences. Questions concerning the removal of wild birds through lethal methods are not uncommon – the removal of great black-backed gulls for conservation, buzzards to preserve air safety, grey herons to protect fisheries – but the killing of robins seemed to provoke especial incredulity: what threat could a robin possibly pose? These questions were reminiscent of a previous story that *The Telegraph* ran in May 2004 on the 'assassination' of three robins at a garden centre in Gloucestershire. The robins were accused of posing a threat to public health by flying in and out of the cafeteria through the air vents and, on one occasion, triggering the security alarm in the early hours of the morning. Given the public outcry that ensued, it is perhaps unsurprising that DEFRA (the Department for Environment, Food and Rural Affairs) was expecting *The Telegraph*'s call and were prepared with a statement concerning the due consideration that was given to the application to remove the birds. This didn't convince the sceptics, who were described as 'aghast' at the news that 'the pitter-patter of tiny robin feet' had been extinguished from the centre's cafeteria in such a 'heavy-handed way'.[16] The article claimed to finally have the answer to who killed Cock Robin: it was a garden centre conspiracy.

The significance of charisma (or lack of it) was made apparent in 2020 when a rufous scrub robin appeared in Norfolk, England, for the first time in forty years. A medium-sized slender bird with a long tail, it is usually found in southern Europe and northeast Africa during the breeding season and south of the Sahara in winter. The scrub robin drew crowds of eager birdwatchers to the north Norfolk coast and made national news, with one birder describing the scrub robin as a little bird with big character. However, one news outlet was less than impressed: 'So is THAT what all the fuss was about?' Alongside a photo of the bird, the article noted that, 'Unlike other exotic birds with their vibrant plumage and striking songs, this one just has pale brown feathers and a thrush-like trill described as having a sad tone.'[17] There have only ever been nine recorded sightings of the species in Britain and Ireland. The most famous sighting of the bird was at a Butlin's holiday camp in Skegness in 1963. As Dick Filby, who runs the information service *Rare Bird Alert*, pointed out, at that time 'people learned about

Rufous scrub robin (*Cercotrichas galactotes*), Seville, Spain.

it by postcard', but now it is only a matter of seconds for news of a rare sighting to travel far and wide.[18]

If our relationship with the natural world is shaped by charisma, then it is worth considering what is at stake for those species that don't seem to possess it. Take the birds that exist at the margins of the robin's story, such as the poor sparrow, whose own cultural history overlaps with that of the robin. This unfortunate bird appears to have been cast as the 'anti-robin' of folk tradition on several occasions. It is for this reason that the robin makes a (not insignificant) appearance in Kim Todd's *Sparrow*. While some stories position the robin as a comfort to Christ in his final hours, the sparrow, meanwhile, is said to have played a very different role, returning the crucifixion nails after swallows had removed them. The sparrow was also said to have prolonged the torture of Christ by alerting the authorities to his continued breathing.[19] The story of Cock Robin is, of course, another slight on the sparrow's character, adding further weight to descriptions of the sparrow's ruthless, downright 'devilish' nature. It would appear that the robin and the sparrow have long been caught in a good bird/bad bird drama, out of which the robin does very well. And, if the sparrow suffers from derisory comparisons with the gardener's friend, it is not helped by its 'successful' introduction to North America. As Robert Fletcher put it in his talk to the Anthropological Society of Washington in 1888, 'The Robin Redbreast is a dainty little bird, scarcely weighing as much as the sparrow, but not at all resembling in shape that detestable little imported pirate.'[20]

The significance of charisma to issues of conservation concern is nothing new, but the robin is unusual in that it fronts several campaigns despite being a species of least conservation concern. This was explicitly acknowledged by the Czech Society for Ornithology in 2016, when it selected the robin to be the Czech bird of the year as an explicit reminder that common bird species also face a variety of threats, including the domestic cat. The point about relatively 'common' species is worth underlining. Conservation science tends to focus attention on scarcity and absence, but there are, of course, a variety of issues that affect even the most abundant of species.

One such issue that the Czech society highlighted during the year of the robin was instances of collision in cities increasingly built of glass. Bird-window collisions in cities are gaining attention – especially in the cities on key migratory routes. For instance, scientific studies recently identified Chicago as one of the most dangerous metropolitan areas in the USA for migratory birds. Situated on one of North America's busiest flyways, Chicago is a city made of towering glass structures. The trend for glass facades has a deadly cost – indeed, conservative estimates suggest that in North America alone a billion birds die every year as a result of window collisions.[21] While slow, new collaborations and building regulations are starting to promote bird-safe lighting and bird-friendly building design. The use of screens, or films and markers on windows, can reduce transparency and the reflectivity of glass, while reducing the number of lights left on overnight can dramatically decrease fatal collisions, as lights are known to distract and disorientate birds.[22] In the USA, a number of cities, including New York, have passed legislation that requires all new constructions to use bird-friendly glass. It would seem that the 'bird-safe' movement is growing, and in 2019, BirdLife International's children's programme, Spring Alive, ran a number of events across Europe and Central Asia to get young people involved in making windows bird-safe, including their own school windows.[23]

To understand the other challenges that the European robin faces, we need look no further than the garden. As biologist and garden enthusiast Dave Goulson has argued, the fate of humankind

is intertwined with the fate of slugs, earwigs, hoverflies, worms and the rest, and so too, I might add, is the fate of robins.[24] Given the robin's diet, healthy soils, abundant insect life and vegetation that can support berry crops are important, as are places to nest. Invertebrate organizations have frequently highlighted the urgency of rapidly declining insect numbers, which are driven by habitat fragmentation, pesticides and climate change. The numbers are stark but repeatedly fail to gain the kind of attention warranted.

A robin fatality after a window collision.

European robin in the Netherlands.

Calls for a move away from overly manicured gardens and road verges in favour of more 'natural', less 'tidy-looking' environments that can better sustain life are starting to gain traction.[25] Pesticides, soil degradation, increased paving and leaf blowers can all have catastrophic effects on insects, which are important not only to pollination but to feeding birds. In 2017 a German report recorded a 76 per cent decline in flying insect numbers across the country.[26] What was dubbed 'Insect Armageddon' in the press prompted the German federal government to announce $118.5 million in funding for insect conservation and issue a statement on leaf blowers, which were implicated in the report. While not going so far as to impose a ban, the government stated that they should only be used when completely necessary, because of their fatal impact on insect life.

As an ambassador for songbirds threatened by a host of ills and ecological crises, the robin's ability to elicit public attention is indisputable. Given the robin's place in folklore, its familiar appeal and wide-ranging popularity, it is unsurprising that it has

Illustration of a robin from John Gerrard Keulemans, *Onze vogels in huis en tuin* (1869).

become a representative for a variety of environmental causes. For centuries, the robin has shaped the way people mourn and how they celebrate Christmas; it has played a role in the stories that we tell ourselves about community, charity and belonging. It has a history that is entangled with imperial endeavours and the whims of consumer culture, while scrutiny of its behaviour has changed scientific understanding of bird migration and ornithological study. At once ordinary and extraordinary, the robin's association with the quality of redness and all that it announces has seen this small bird take on a wealth of meaning. But while debates over the robin's future and character persist, the robin continues, unconcerned, with its melodious and lilting whistles.

Timeline of the Robin

c. 32 MYA

The earliest known songbirds in Europe

c. 350 BC

Aristotle writes his theory of transmutation, suggesting that robins and redstarts change into one another

AD 600

St Mungo brings a robin back to life. The robin will become part of the founding story of the city of Glasgow as a result

1595

The Children in the Wood is published as a ballad

1786

Sarah Trimmer's *Fabulous Histories* is published. It is later renamed *History of the Robins*

1800

The robin is reassigned to the genus *Erithacus* by Georges Cuvier

1843

The first Christmas card is sent

1965–6

Merkel and Wiltschko publish their findings on robins and geomagnetism – a breakthrough in the understanding of bird migration

1962

Rachel Carson's *Silent Spring* is published

1964

An American robin erroneously appears in Walt Disney's *Mary Poppins*

1695

A robin dubbed the 'Westminster Wonder' appears at Westminster Abbey while the body of Queen Mary II lies in state

1744

'Who Killed Cock Robin?' is published in *Tommy Thumb's Pretty Song Book*

1758

The robin appears in the 10th edition of Linnaeus's *Systema Naturae* as *Motacilla rubecula*

1876

The 'Robin Dinner' movement is established

1890

Ogden launches Robin cigarettes and Redbreast Flakes

1911

Frances Hodgson Burnett's *The Secret Garden* is published

1943

David Lack's ground-breaking study, *The Life of the Robin*, is published

1979

The Bern Convention makes the trapping and killing of wild birds illegal

2010

After genetic testing, the European robin becomes the sole member of the genus *Erithacus*

2016

The robin is voted Britain's favourite bird in a public poll

References

1 A Familiar Bird

1 Kyle Mullin, 'Rockin' Robin: Beijingers Flock to Spot Rare Bird', *The Beijinger*, www.thebeijinger.com, 10 January 2019.

2 David Lack et al., *The Life of the Robin* (London, 2016), frontmatter.

3 J. P. Burkitt, 'A Study of the Robin by Means of Marked Birds', *British Birds* (1924–6), 17, pp. 294–303; 18, pp. 97–103, 250–57; 19, pp. 120–24; 20, pp. 91–101.

4 J. Van Tyne, 'The Life of the Robin', *Wilson Bulletin*, 55 (1943), pp. 252–6.

5 T. Anderson, *The Life of David Lack: Father of Evolutionary Ecology* (Oxford, 2013), p. 35.

6 Andrew Lack, *Redbreast: The Robin in Life and Literature* (Pulborough, 2008).

7 See also Chris Mead, *Robins* (London, 1984); Stephen Moss, *The Robin: A Biography* (London, 2017).

8 Peter Clement and Chris Rose, *Robins and Chats* (London, 2015), pp. 244–5.

9 Marianne Taylor, *Robins* (London, 2015), p. 9; 'The Round Robin: Roly-Poly Bird Who Looks Like He's Swallowed a Christmas Bauble', *Daily Mail* (14 December 2010).

10 Clement and Rose, *Robins and Chats*, p. 241.

11 Ibid.

12 Frances Hodgson Burnett, *The Secret Garden* [1911] (London, 1983), p. 52.

13 Frances Hodgson Burnett, *My Robin* (New York, 1912), pp. 1–10.

14 Ibid., pp. 16–17.
15 G. Gerzina, *Frances Hodgson Burnett: The Unpredictable Life of the Author of 'The Secret Garden'* (London, 2005), p. 268.
16 L. Lear, *Beatrix Potter: A Life in Nature* (London, 2007).
17 M. McDowell, *Beatrix Potter's Gardening Life: The Plants and Places that Inspired the Classic Children's Tales* (London, 2013), p. 182.
18 G. W. Temperley, 'Individuality in Birds: A Study of Robins' (Northumbria, 1961), available in the Natural History Society Northumbria archives.
19 Kate McRae, 'Bird on the Wire', *The Times* (15 May 2020). See also www.wildlifekate.co.uk, 19 May 2020.

2 THE GLOBAL ROBIN

1 W. E. Boles, 'Fossil Songbirds (Passeriformes) from the Early Eocene of Australia', *Emu*, XCVII/1 (1997), pp. 43–50; R. G. Moyle et al., 'Tectonic Collision and Uplift of Wallacea Triggered the Global Songbird Radiation', *Nature Communications*, VII (2016), pp. 1–7.
2 Z. M. Bochenski et al., 'A New Passeriform (Aves: Passeriformes) from the Early Oligocene of Poland Sheds Light on the Beginnings of Suboscines', *Journal of Ornithology*, CLXII (2021), pp. 1–12.
3 See Peter Clement and Chris Rose, *Robins and Chats* (London, 2015).
4 G. Sangster et al., 'Multi-Locus Phylogenetic Analysis of Old World chats and Flycatchers Reveals Extensive Paraphyly at Family, Subfamily and Genus Level (Aves: uscicapidae)', *Molecular Phylogenetics and Evolution*, LXXII/1 (2010), pp. 380–92.
5 See W. Blunt, *Linnaeus: The Compleat Naturalist* (Princeton, NJ, 2001).
6 Marianne Taylor, *Robins* (London, 2015), p. 20.
7 For example, it appears in Chaucer's *The Parliament of Fowls* in the late fourteenth century.
8 See Spike Bucklow, *Red: The Art and Science of a Colour* (London, 2016), p. 12.
9 William Shakespeare, *Cymbeline; A Tragedy, with the notes and illustrations of various commentators* (London, 1794).
10 Stephen Moss, *The Robin: A Biography* (London, 2017), p. 27.

11 E. G. Withycombe, *The Oxford Dictionary of English Christian Names* (Oxford, 1977).

12 Thomas Bewick, *History of British Birds* (London, 1797), p. 207.

13 William Wordsworth, 'The Redbreast and the Butterfly', in *Poems* (London, 1815), vol. i, pp. 261–2 (the title of the poem was changed in 1845 to 'The Redbreast Chasing the Butterfly').

14 H. Dunn and J. T. Crossley, *Daily Lesson Book for the Use of Schools and Families, no. iii* (Hobart, 1849).

15 A. S. Palmer, *Folk-Etymology: A Dictionary of Verbal Corruptions or Words Perverted in Form or Meaning* (London, 1882), p. 283.

16 Mary Louise Pratt, *Imperial Eyes: Travel Writing and Transculturation* (London, 2008).

17 J. Byron et al., *An Account of the Voyages Undertaken by the Order of His Present Majesty, for Making Discoveries in the Southern Hemisphere: And Successively Performed by Commodore Byron, Captain Wallis, Captain Carteret, and Captain Cook, in the Dolphin, the Swallow, and the Endeavour: Drawn Up from the Journals which Were Kept by the Several Commanders, and from the Papers of Joseph Banks* (London, 1775), vol. ii, p. 242.

18 J. C. Beaglehole, ed., *The Journals of Captain James Cook on his Voyages of Discovery*, vol. ii: *The Voyage of the Resolution and Adventure, 1772–1775* (Cambridge, 1961), p. 424.

19 J. J. Audubon, *The Birds of America, from Drawings Made in the United States* (London, 2006), p. 496.

20 O. L. Austin, *Birds of the World* (New York, 1961), p. 255.

21 Scarlet-breasted robin (*Petroica boodang*), in J. Gould's *Birds of Australia* (London, 1848), vol. iii, p. 3. Also known as red-breasted warbler.

22 Gould, *Birds of Australia*, p. 5.

23 Peter Menkhorst et al., *The Australian Bird Guide* (Princeton, NJ, and Woodstock, 2017), pp. 470–742.

24 T. R. Dunlap, 'Remaking the Land: The Acclimatization Movement and Anglo Ideas of Nature', *Journal of World History*, viii/2 (1997), pp. 303–19; R. Flikke, 'Domestication of Air, Scent, and Disease', in *Domestication Gone Wild: Politics and Practices of Multispecies Relations*, ed. H. A. Swanson et al. (Durham, NC, 2019), pp. 176–95.

25 Dunlap, 'Remaking the Land', p. 306.
26 Gilbert White, *The Natural History of Selborne* [1789] (Oxford, 2013), p. 83.
27 Dunlap, 'Remaking the Land', p. 307.
28 D. Tout-Smith, 'Acclimatisation Society of Victoria', in Museums Victoria Collections, https://collections.museumsvictoria.com.au, accessed 25 February 2019; L. Gillbank, 'The Origins of the Acclimatisation Society of Victoria: Practical Science in the Wake of the Gold Rush', *Historical Records of Australian Science* (1984), pp. 359–74.
29 '"If It Lives, We Want It." Exploring the Acclimatisation Society of Victoria's Role in Australia's Ecological History', https://blog.biodiversitylibrary.org, accessed 2 May 2021.
30 'Annual Report of the Acclimatisation Society of Victoria: With the Addresses Delivered at the Annual Meeting of the Society' (1864), vol. III, p. 33, www.biodiversitylibrary.org, accessed 3 September 2021.
31 'Acclimatisation', *Press*, I/13 (17 August 1861), p. 1, https://paperspast.natlib.govt.nz, accessed 3 September 2021.
32 J. C. Phillips, *Wild Birds Introduced or Transplanted in North America* (Washington, DC, 1928).
33 There is a third, ambiguous, reference to a 'redbreast' in *Henry IV*, but there is insufficient clarity to suggest that the intended bird was not in fact a bullfinch. See J. E. Harting, *The Ornithology of Shakespeare* (London, 1871), pp. 140–42.
34 P. Kalm, *Travels into North America* (London, 1771), vol. II, p. 90.
35 Austin, *Birds of the World*, p. 251.
36 Rachel Carson, *Silent Spring* [1962] (London, 2000).
37 Ibid., p. 101.
38 Ibid., p. 109.
39 Austin, *Birds of the World*, p. 252.
40 Roland H. Wauer, *The American Robin* (Austin, TX, 1999), p. 1.
41 *Wehman's Cook Book: A Complete Collection of Valuable Recipes Suited to Every Household and All Tastes* (New York, 1890), p. 42.

42 'Migratory Bird Treaty Act', https://fws.gov, accessed 2 May 2021.
43 T. C. Jerdon, *The Birds of India*, vol. II, part 1 (Calcutta, 1863), p. 116.
44 Clement and Rose, *Robins and Chats*, pp. 384–8.
45 A. Newton, ed., *A Dictionary of Birds* (London, 1893–6), p. 133.
46 E. Albin, *A Supplement to the Natural History of Birds*, 17 (London, 1737).
47 F. Levaillant, *Histoire naturelle des oiseaux d'Afrique* (Paris, 1802), vol. III, pp. 50–52.
48 'A moins qu'on ne l'attribue à ce que, chantant à certaines heures du jour, comme il le fait en effet, les peuples de cette partie du monde, où l'espèce se trouve aussi bien qu'en Afrique, ne soient instruits de l'heure qu'il est par le moment où le mâle commence son chant; ce qui est assez ordinairement au lever et au coucher du soleil', ibid., p. 50.
49 Levaillant, *Histoire naturelle*, p. 50.
50 Jerdon, *The Birds of India*, p. 116.
51 E. Blyth, 'The Ornithology of India: A Commentary on Dr Jerdon's Birds of India', *Ibis*, IX/1 (1867), pp. 147–85.
52 A. Newton, ed., *A Dictionary of Birds* (London, 1896), p. 133.
53 Jerdon, *The Birds of India*, p. 116.
54 J. Gould, *The Birds of Asia* (London, 1850–83), vol. III, pp. 26–7.
55 E. Layard ibid., p. 26.
56 Jerdon, *The Birds of India*, p. 116.
57 Ibid.
58 G. Sangster et al., 'Multi-locus Phylogenetic Analysis of Old World Chats and Flycatchers Reveals Extensive Paraphyly at Family, Subfamily and Genus Level (Aves: Muscicapidae)' *Molecular Phylogenetics and Evolution*, LVII/1 (2010), pp. 380–92.

3 DEATH, BAD OMENS AND A PIOUS BIRD

1 Canon Kent, *The Land of the 'Babes in the Wood'; or, the Breckland of Norfolk*. See Andrew Lack, *Redbreast: The Robin in Life and Literature* (Pulborough, 2008), pp. 50–77 for a discussion of the ballad's origins.

2 A recusant was a person who refused to attend the services of the Church of England.

3 'The Children in the Wood', in Joseph Jacobs, *More English Fairy Tales* (New York and London, 1894), pp. 111–17.

4 Thomas Crawford, *The Babes in the Wood* (*c.* 1850), www.metmuseum.org, accessed 2 May 2021.

5 The Babes in the Wood, www.metmuseum.org, accessed 3 September 2021.

6 James Russell Lowell, 'The First Snowfall', in *Poems* (Cambridge, 1848).

7 Roland H. Wauer, *The American Robin* (Austin, TX, 1999), pp. 10–11.

8 S. Finch, *Wordsworth's Birds* (Lancaster, 1986), p. 145. Finch's volume erroneously refers to Fort Fuentes as 'Fort Fluentes'.

9 William Wordsworth, *Memorials of a Tour on the Continent* (London, 1822), lines 1–4 and 9–12, p. 31.

10 *The Pitt Press Shakespeare: Cymbeline*, ed. A. W. Verity (Cambidge, 1960), p. 157.

11 'The Westminster Wonder:/ Giving an Account of a Robin Red-breast, who, ever since the Queen's/ Funeral, continues on the top Pinacle of the Queen's Mausoleum, or Pyramid,/ in the Abby of Westminster, where he is seen and heard to sing, and will not/ depart the place, to the admiration of many Beholders' (London, 1695), printed for J. Blare, at the Sign of the Looking-glass on London-bridge. Collection: Magdalene College – Pepys.

12 *On a robin red-breast, who took up his residence in Bristol Cathedral, and accompanied the organ with his singing*, printed by John Fowler [1785], Eighteenth Century Collections online, www.gale.com, accessed 3 September 2021.

13 'The Robin Redbreast of the Bristol Cathedral', *Sherborne Mercury* (24 April 1860), p. 7.

14 William Wordsworth, 'To a Redbreast', in *Miscellaneous Poems* (1842).

15 Niall Mac Coitir, *Ireland's Birds: Myths, Legends and Folklore* (Cork, 2015), pp. 17–24.

16 C. L. Daniels and C. M. Stevans, eds, *Encyclopædia of Superstitions, Folklore, and the Occult Sciences of the World*, vol. II [1903] (Honolulu, HI, 2003), p. 687.
17 G. P. Olina, *Pasta for Nightingales: A 17th-Century Handbook of Bird-Care and Folklore* (New Haven, CT, 2018), p. 12.
18 Niall Mac Coitir, *Ireland's Birds*, p. 116.
19 Daniels and Stevans, eds, *Encyclopædia of Superstitions*, p. 687.
20 W. Howitt, *The Three Death-Cries of a Perishing Church* (Nottingham, 1835), pp. 30–32.
21 Ibid.
22 J. Cahuac, *Who Killed Cock Robin? A Satirical Tragedy, or Hieroglyphic Prophecy on the Manchester Blot!!!* (London, 1819).
23 Robert Reid, *The Peterloo Massacre* (London, 2017).
24 P. Morris and J. Ebenstein, *Walter Potter's Curious World of Taxidermy* (London, 2013), p. 2.
25 C. Haddon, 'Sad Fate for a Unique Collection', *Sunday Times* (1 July 1973), p. 13.
26 N. Brown, *Fairies in Nineteenth-Century Art and Literature* (Cambridge, 2001), vol. XXXIII.
27 S. Stewart, *On Longing: Narratives of the Miniature, the Gigantic, the Souvenir, the Collection* (Durham, NC, 1993), p. 112.
28 Brown, *Fairies*, p. 39.
29 Walt Disney, *Who Killed Cock Robin?*, release date 26 June 1935, dir. David Hand.
30 R. Merritt and J. B. Kaufman, *Walt Disney's Silly Symphonies: A Companion to the Classic Cartoon Series* (Gemona, Udine, 2006), p. 23; M. Barrier, *Hollywood Cartoons: American Animation in Its Golden Age* (Oxford, 2003), p. 135.
31 R. B. Hale, *The Beloved St Mungo, Founder of Glasgow* (Ottawa, 1989), p. 16.
32 Ibid., p. 18.
33 Ted R. Anderson, *The Life of David Lack: Father of Evolutionary Ecology* (Oxford, 2013), p. 27.
34 Peter Clement and Chris Rose, *Robins and Chats* (London, 2015), p. 24.

4 A BIRD OF SONG AND SEASONAL CHANGE

1 A. Huxley cited in J. Huxley, 'Preface', in Rachel Carson, *Silent Spring* [1962] (London, 2000), p. 20.
2 Peter Clement and Chris Rose, *Robins and Chats* (London, 2015), p. 240.
3 Ibid.
4 E. Grey, *The Charm of Birds* (New York, 1927), p. 8.
5 'Robin Redbreast Vices', *Aberdeen Journal* (8 January 1932), p. 6.
6 Andrew Lack, *Redbreast: The Robin in Life and Literature* (Pulborough, 2008), p. 90.
7 Andrew Motion, *Keats* (London, 1997), p. 461.
8 Martin Middeke and Christina Wald, 'Melancholia as a Sense of Loss: An Introduction', in *The Literature of Melancholia: Early Modern to Postmodern* (Basingstoke, 2011), p. 1.
9 S. Finch, *Wordsworth's Birds* (Lancaster, 1986), p. 146.
10 Stephen Hebron, 'An Introduction to "Ode on Melancholy"', www.bl.uk, 15 May 2014.
11 William Wordsworth 'Book Seventh: Residence in London', quoted in Finch, *Wordsworth's Birds*, pp. 142–3.
12 David Hill, *Turner's Birds: Bird Studies and Landscapes from Farnley Hall* (Oxford, 1988), p. 9.
13 Ibid., p. 18.
14 William Warner Caldwell, 'Robin's Come', *Poems: Original and Translated* (Boston, MA, and Cambridge, 1857), p. 7.
15 John Burroughs, *Wake-Robin* (Boston, 1887), vol. I, pp. 9–10.
16 'Robin Snow', www.oed.com, accessed 3 September 2021.
17 W. H. Bergtold, 'Intoxicated Robins', *The Auk*, XLVII/4 (1 October 1930), p. 571.
18 L. L. Haupt, *Crow Planet: Essential Wisdom from the Urban Wilderness* (New York, 2009), p. 74.
19 G. P. Olina, *Pasta for Nightingales: A 17th-Century Handbook of Bird-care and Folklore* (New Haven, CT, 2018), p. 13.
20 Nicole Sault, 'Bird Messengers For All Seasons: Landscapes of Knowledge Among the Bribri of Costa Rica', in *Ethno-Ornithology: Birds, Indigenous Peoples, Culture and Society*,

ed. Sonia C. Tidemann and Andrew Gosler (London and Washington, DC, 2010), p. 292.

21 Ibid., p. 297.

22 Agatha, *De roodborstjes* (Amsterdam, 1874).

23 Ibid., p. 6.

24 D. Hoogenboezem, 'Marvel, Feminism and Reason: Rewriting Marie-Catherine d'Aulnoy's Fairy Tales for Dutch Children', in *Readers, Writers, Salonnières: Female Networks in Europe, 1700–1900*, ed. H. Brown and G. Dow (Bern and Oxford, 2011), pp. 259–76.

25 David Lack et al., *The Life of the Robin* (London, 2016), p. 10.

26 Mark Cocker and Richard Mabey, *Birds Britannica* (London, 2005), p. 337.

27 Clement and Rose, *Robins and Chats*, p. 385.

28 K. A. Rentschlar et al., 'A Silent Morning: The Songbird Trade in Kalimantan, Indonesia', *Tropical Conservation Science*, 11 (2018), pp. 1–10.

29 IUCN SSC Asian Songbird Trade Specialist Group, www.asiansongbirdtradesg.com, accessed 14 September 2021.

30 Z. Burivalova et al., 'Understanding Consumer Preferences and Demography in Order to Reduce the Domestic Trade in Wild-Caught Birds', *Biological Conservation*, 209 (2017), pp. 423–31.

31 A. Kirichot, S. Untaya and S. Singyabuth, 'The Culture of Sound: A Case Study of Birdsong Competition in Chana District, Thailand', *Asian Culture and History*, VII/1 (2015), p. 5.

32 Paul Jepson, 'Orange-Headed Thrush *Zoothera citrina* and the Avian X-Factor', *Birding Asia*, IX (2008), pp. 58–61.

33 Paul Jepson, 'Towards an Indonesian Bird Conservation Ethos: Reflections from a Study of Bird-Keeping in the Cities of Java and Bali', in *Ethno-Ornithology*, ed. Tidemann and Gosler, p. 322.

34 Merle Patchett, 'The Biogeographies of the Blue Bird-of-Paradise: From Sexual Selection to Sex and the City', *Journal of Social History*, LII/4 (2019), pp. 1061–86.

35 Lydia Slater, 'Practically Perfect: Emily Blunt on Motherhood, Magic and Taking On Mary Poppins', *Harper's Bazaar*, www.harpersbazaar.com, 28 November 2018.

36 Peter Cashwell, 'Why That American Robin Cameo in "The Hobbit" Wasn't an Error', www.audubon.org, 14 October 2016.
37 Sean Adams, *The Designer's Dictionary of Color* (New York, 2017), p. 181.
38 P. A. English and R. Montgomerie, 'Robin's Egg Blue: Does Egg Color Influence Male Parental Care?', *Behavioral Ecology and Sociobiology*, LXV/5 (2011), pp. 1029–36.
39 Chris Mead, *Robins* (London, 1984), p. 64.
40 Clement and Rose, *Robins and Chats*, pp. 241–2.
41 Lack et al., *The Life of the Robin*, p. 83.
42 E. Leigh, *Ballads and Legends of Cheshire* (London, 1867), p. 61. The rhyme also features in Andrew Lack's *Redbreast* in a chapter on the relationship between robins and prisoners.
43 W. Thompson, *The Natural History of Ireland* (London, 1849), vol. I, p. 163.
44 'A Curious Nest Site', *Western Daily Press* (20 May 1930), p. 7; 'Robins' Nest in Crematorium', *Western Times* (14 April 1927), p. 12; 'Robins Nest in Letter-Box', *Derby Daily Telegraph* (24 May 1938), p. 6; 'Robin Builds Nest on Engineering Bench', *Evening Telegraph* (1 May 1935), p. 8; 'Robins Nest in Car', *Daily Telegraph* (17 April 1957), p. 9; 'Robin's Nest on Books', *Daily Mail* (7 May 1913), p. 5; 'Robin Nests in Workman's Haversack', *Gloucester Citizen* (13 April 1950), p. 5.
45 Roshini Muthukumar, 'TN Village Turned Off Street Lights for Over a Month to Protect a Bird's Home', www.thebetterindia.com, 23 July 2020. In several reports it was said to be an Indian robin.

5 FOR TERRITORY AND NATION

1 M. Jay, 'Timbremelancholy: Walter Benjamin and the Fate of Philately', *Salmagundi*, 194 (2017), p. 33; Walter Benjamin, *One-Way Street and Other Writings*, new edn (London, 2009), pp. 100–103.
2 'Mission', www.romfilatelia.ro, accessed 3 September 2021.
3 Jay, 'Timbremelancholy'.
4 J. Broom, *A History of Cigarette and Trade Cards: The Magic Inside the Packet* (Barnsley, 2018).

5 Robin, Ogden's British Birds Series UK Series, 1909.
6 D. Wainwright, *Brooke Bond: A Hundred Years* (London, 1970).
7 Broom, *A History of Cigarette and Trade Cards*.
8 David Lindo, 'Introduction', in David Lack et al., *The Life of the Robin* (London, 2016), p. xiv.
9 J. Stamp, 'American Myths: Benjamin Franklin's Turkey and the Presidential Seal', www.smithsonianmag.com, 25 January 2013.
10 'Getting to Know Michigan', www.legislature.mi.gov, accessed 6 May 2021.
11 'The State Bird', https://portal.ct.gov, accessed 6 May 2021; 'Wisconsin State Bird', www.netstate.com, accessed 6 May 2021.
12 S. Bhagat, 'In a First, City to Get "Bird of Mumbai"', *Times of India* (22 February 2011).
13 Lindo, 'Introduction', p. xiii.
14 Philip Hoare, 'Britain Has Spoken – and Chosen a Vicious Murdering Bully as Its National Bird', *The Guardian* www.theguardian.com, 11 June 2015.
15 Anna van Praagh, 'Is the Robin Too Ordinary to Be Britain's National Bird?', *The Telegraph*, www.telegraph.co.uk, 16 March 2015.
16 David Lack, 'Early References to Territory in Bird Life', *The Condor*, XLVI/3 (1944), pp. 108–11.
17 G. W. Temperley, 'Individuality in Birds: A Study of Robins' (Northumbria, 1961).
18 Frances Hodgson Burnett, *My Robin* (New York, 1912), p. 43.
19 Marianne Taylor, *Robins* (London, 2015), p. 49.
20 William Thompson, *The Natural History of Ireland*, vol. I: *Birds, Comprising the Orders Raptores and Insessores* (London, 1849), pp. 64–6.
21 Ibid.
22 Taylor, *Robins*, p. 48.
23 David Lack et al., *The Life of the Robin* (London, 2016), p. 59.
24 Ross Hoddinott, *Territorial Strut*, www.nhm.ac.uk, accessed 6 May 2021.
25 Peter Clement and Chris Rose, *Robins and Chats* (London, 2015), p. 244.

26 Ibid.
27 Aristotle, *The History of Animals*, Book 9 chapter 49B.
28 Lisa Pollack, 'That Nest of Wires We Call the Imagination: A History of Some Key Scientists Behind the Bird Compass Sense', www.ks.uiuc.edu, May 2012.
29 Tim Birkhead, *Bird Sense: What It's Like to Be a Bird* (London, 2012), p. 174.

6 THE COLOUR RED AND A CHRISTMAS STORY

1 F. A. Fulcher, 'Robin Blue Breast', *English Illustrated Magazine*, 45 (December 1906), p. 237.
2 Michel Pastoureau, *Red: The History of a Color*, trans. Jody Gladding (Princeton, NJ, and Woodstock, 2016), p. 22.
3 Spike Bucklow, *Red: The Art and Science of a Colour* (London, 2016), p. 21.
4 Selma Lagerlöf, *Christ Legends* (New York, 1908), pp. 91–202.
5 Jo Nesbø, *The Redbreast* (London, 2006), frontmatter.
6 Dandi Daley Mackall, *The Legend of the Easter Robin: An Easter Story of Compassion and Faith* (Grand Rapids, MI, 2016).
7 G. R. Fulton, 'The Scarlet Robin's Red Breast: An Indigenous Narrative', *Australian Field Ornithology*, XXII/4 (January 2005), p. 213; 'Robin's Red Breast a Badge of Courage', *Sunday Morning Herald* (29 March 2014).
8 'How Tol-le-loo Stole Fire: A Miwok Legend', www.firstpeople.us, accessed 30 May 2021.
9 'Nukumi and Fire', www.native-languages.org, accessed 30 May 2021.
10 Robert Fletcher, 'Myths of the Robin Redbreast in Early English Poetry', *American Anthropologist*, II/2 (1889), p. 113.
11 William Edward Hartpole Lecky, *History of European Morals* (London, 1869), p. 221.
12 C. S. Lewis, *The Chronicles of Narnia* (London, 1998), pp. 137–9.
13 Marianne Taylor, *Robins* (London, 2015), p. 13.
14 'Club History', www.swindontownfc.co.uk, 1 April 2017.

15 Anthony Trollope, *Framley Parsonage* [1861] (London, 2016), pp. 54–6.
16 Anthony Trollope, *An Autobiography* (New York, 1883), pp. 31–52.
17 See, for example, Alison Uttley, *The Squirrel, the Hare and the Little Grey Rabbit* (London, 1929), p. 13.
18 'Pop It in the Post. How the Penny Black Stamp Changed Our World', www.postalmuseum.org, accessed 30 May 2021.
19 Patricia Zakreski, 'The Victorian Christmas Card as Aesthetic Object: Very Interesting *Ephemeræ* of a Very Interesting Period in English Art-Production', *Journal of Design History*, XXIX/2 (May 2016), pp. 120–36.
20 Neil Armstrong, *Christmas in Nineteenth-Century England* (Manchester, 2010), p. 94.
21 Stephen Moss, *The Wren: A Biography* (London, 2018).
22 John Grossman, *Christmas Curiosities: Odd, Dark and Forgotten Christmas* (New York, 2008).
23 Armstrong, *Christmas in Nineteenth-Century England*, p. 105.
24 Ibid.
25 C. Bullock, *Robin's Carol, and What Came of It: The Story of Robin Dinners* (London, 1879), p. 22.
26 Ibid.
27 Pat Wynnejones, *The Story of Robin Redbreast*, retold from Mrs Alfred Gatty's *Parables From Nature* (Singapore, 1989). No pagination.
28 Sarah Trimmer, *History of the Robins*, 2nd edn (Dublin, 1821), p. A3.
29 Ibid.
30 Wilfried Keutsch, 'Teaching the Poor: Sarah Trimmer, God's Own Handmaid', *Bulletin of the John Rylands Library*, LXXVI/3 (1994), pp. 43–57.
31 Linda Lear, *Beatrix Potter: A Life in Nature* (London, 2008), p. 34.
32 Pastoureau, *Red*, p. 135.
33 Peter Clement and Chris Rose, *Robins and Chats* (London, 2015), p. 254.
34 Ibid., pp. 235–8.

7 ROBIN FUTURES

1 A. J. Burt et al., 'The History, Status and Trends of the Endangered Seychelles Magpie-Robin *Copsychus sechellarum*', *Bird Conservation International*, xxvi/4 (2016), pp. 505–23.

2 P. Armstrong and A. Potts, 'The Emptiness of the Wild', in *Routledge Handbook of Human-Animal Studies*, ed. G. Marvin and S. McHugh (London, 2014), pp. 168–81.

3 A. da Silva et al., 'Artificial Night Lighting Rather Than Traffic Noise Affects the Daily Timing of Dawn and Dusk Singing in Common European Songbirds', *Behavioral Ecology*, xxv/5 (2014), pp. 1037–47.

4 M. W. Miller, 'Apparent Effects of Light Pollution on Singing Behavior of American Robins', *The Condor*, cviii/1 (2006), pp. 13–139.

5 Da Silva et al. 'Artificial Night Lighting', p. 8.

6 M. C. Zwart, 'Wind Farm Noise Suppresses Territorial Defense Behavior in a Songbird', *Behavioral Ecology*, xxvii/1 (2016), pp. 101–8.

7 M. Whittingham, 'Robin Hushed: Wind Turbines are Making Songbirds Change Their Tune', www.theconversation.com, 20 December 2018.

8 Convention on the Conservation of European Wildlife and Natural Habitats, www.coe.int, accessed 6 May 2021; European Commission, Directorate-General for Environment, 'The Birds Directive: 40 Years of Conserving Our Shared Natural Heritage' (2019).

9 S. Bhattacharya, 'Slaughter of the Song Birds', *Nature*, dxxix/7587 (2016), pp. 452–6.

10 H. M. Jenkins, C Mammides and A. Keane, 'Exploring Differences in Stakeholders' Perceptions of Illegal Bird Trapping in Cyprus', *Journal of Ethnobiology and Ethnomedicine*, xiii/1 (2017), pp. 1–10.

11 C. Sellers, *Tales from the Lands of Nuts and Grapes: Spanish and Portuguese Folklore* (London, 1888), pp. 112–16.

12 See www.thelasttweet.eu, accessed 6 May 2021.

13 Stuart Winter, 'Skinned and Ready to Eat: British Robins Among 25m Birds Slaughtered in Southern Europe', *The Express*, www.express.co.uk, 22 August 2015.
14 Jamie Lorimer, 'Nonhuman Charisma', *Environment and Planning D: Society and Space*, xxv/5 (2007), pp. 911–32.
15 See www.nationalrobinday.co.uk, accessed 6 May 2021.
16 'Who Killed Cock Robin? Actually It Was a Garden Centre Conspiracy', *The Telegraph*, www.telegraph.co.uk, 23 May 2004.
17 Shari Miller, 'So is THAT What All the Fuss Is About?', Mail Online, www.dailymail.co.uk, 18 October 2020.
18 S. Anderson, 'Visitors Flock to the Coast for Sight of Bird Rarely Seen in UK – Despite Police Warning', *Eastern Daily Press* (19 October 2020), p. 5.
19 Kim Todd, *Sparrow* (London, 2012), pp. 80–89.
20 Robert Fletcher, 'Myths of the Robin Redbreast in Early English Poetry', *American Anthropologist*, II/2 (1889), p. 98.
21 D. Klem Jr, 'Bird–Window Collisions: A Critical Animal Welfare and Conservation Issue', *Journal of Applied Animal Welfare Science*, XVIII (2015), pp. S11–S17; J. A. Elmore et al., 'Predicting Bird–Window Collisions with Weather Radar', *Journal of Applied Ecology*, (2021), pp. 1–9.
22 Chicago Bird Collision Monitors, www.birdmonitors.net, accessed 6 May 2021.
23 Jessica Law, '"Bird-Safe Window" Movement Reaches Over 5 Million People', Birdlife International, www.birdlife.org, 3 July 2019.
24 Dave Goulson, *The Garden Jungle; or, Gardening to Save the Planet* (London, 2019).
25 Ibid.
26 G. Vogel, Where Have All the Insects Gone?', *Science* (2017), pp. 576–9; C. A. Hallmann et al., 'More than 75 percent Decline over 27 Years in Total Flying Insect Biomass in Protected Areas', *PLOS One*, 12, e0185809 (2017).

Select Bibliography

Armstrong, Neil, *Christmas in Nineteenth-Century England* (Manchester, 2010)

Beatley, Timothy, *The Bird-Friendly City: Creating Safe Urban Habitats* (Washington, DC, 2020)

Burnett, Frances Hodgson, *The Secret Garden* [1911] (London, 2017)

—, *My Robin* (New York, 1912)

Carson, Rachel, *Silent Spring* [1962] (London, 1999)

Clement, Peter, and Chris Rose, *Robins and Chats* (London, 2015)

Cocker, Mark, and Richard Mabey, *Birds Britannica* (London, 2005)

Gatty, Mrs Alfred, 'Daily Bread', in *Parables From Nature* (London, 1855), pp. 131–50

Goulson, Dave, *The Garden Jungle: Or Gardening to Save the Planet* (London, 2019)

Lack, Andrew, *Redbreast: The Robin in Life and Literature* (Pulborough, 2008)

Lack, David et al., *The Life of the Robin* (London, 2016)

Lagerlöf, Selma, *Christ Legends* (New York, 1908)

Mead, Chris, *Robins* (London, 1984)

Menkhorst, Peter et al., *The Australian Bird Guide* (Princeton, NJ, and Woodstock, 2017)

Moss, Stephen, *The Robin: A Biography* (London, 2017)

Pastoureau, Michel, *Red: The History of a Color*, trans. Jody Gladding (Princeton, NJ, and Woodstock, 2017)

The Sechelt Nation, *How the Robin Got Its Red Breast: A Legend of the Sechelt People*, illus. Charles Craigan (Roberts Creek, BC, 1993)

Taylor, Marianne, *Robins* (London, 2015)

Tidemann, Sonia, and Andrew Gosler, eds, *Ethno-Ornithology: Birds, Indigenous Peoples, Culture and Society* (London and Washington, DC, 2010)
Trimmer, Sarah, *History of the Robins*, 2nd edn (Dublin, 1821)
Wauer, Roland H., *The American Robin* (Austin, TX, 1999)
Wiltschko, Wolfgang, and Roswitha Wiltschko, 'Magnetic Compass of European Robins', *Science*, CLXXVI/4030 (1972), pp. 62–4

Associations and Websites

ASIAN SONGBIRD TRADE SPECIALIST GROUP

www.asiansongbirdtradesg.com

Focused on preventing the extinction of songbirds threatened by the unsustainable trapping and trade of wild birds. The website has information about the group, current research and legislation, as well as educational resources and publication lists.

BIRDLIFE INTERNATIONAL

www.birdlife.org

A global partnership of conservation organizations and a world leader in bird conservation, bird habitats and global biodiversity. Alongside information about their global work, the website has a data zone that will allow you to look up different species, including over a hundred different 'robins'.

BUGLIFE

www.buglife.org.uk

A UK-based non-profit organization focused on the conservation of invertebrates. It raises awareness of the importance of invertebrates and current invertebrate loss and supports conservation initiatives. The webite has lots of information about current projects, why invertebrates are significant to ecosystems and bird life, bug directories and identification tools, and how to get involved in the reversal of insect declines, including tips on wildlife gardening.

CHICAGO BIRD COLLISION MONITORS

www.birdmonitors.net

A volunteer conservation project dedicated to the protection of migratory birds through rescue, advocacy and outreach. The website has lots of information about the team's work and research, details about the threats that urban birds such as the American robin face, and material about bird-safe lighting and building design for the prevention of collision.

NATIVE LANGUAGES OF THE AMERICAS

www.native-language.org

A small, non-profit organization focused on the survival of Native American languages. The website offers an array of online materials about more than eight hundred indigenous languages of the western hemisphere and the people who speak them. It has an excellent set of resources on Native American animal mythology, including the American robin and many other birds.

THE WILDLIFE TRUSTS

www.wildlifetrusts.org/gardening

Independent charities across the UK. The Wildlife Trusts focus on the protection of wildlife, natural solutions to climate change, and environmental education. The website has plenty of information on wildlife gardening including how to attract birds to your garden, support nesting and plant with wildlife value in mind.

Acknowledgements

I owe many thanks to the team at Reaktion. Especial thanks are due to Jonathan Burt for the thought-provoking comments and editorial suggestions, and to Susannah Jayes and Amy Salter for their attention to detail and endless patience and support in getting the book through to production.

I am extremely grateful to the many people that responded to my image requests and thus helped bring *Robin* to life. During the early stages of the book, Hellen Pethers and Paul Cooper from the Library and Archives team at the Natural History Museum, London, were incredibly helpful.

I would like to thank the brilliant friends that have supported me while writing *Robin*, including Leah Gibbs and Anna Secor, with particular thanks to Ben Anderson for *The Lion, The Witch and the Wardrobe* references and his unwavering enthusiasm for the project.

Thanks are due to Martin and Kathryn Miller, with especial thanks to Kathryn for her continued friendship and support, and for supplying me with numerous robin trinkets and baubles. Many thanks to Sophie Miller for her excellent orange giraffe or what came to be known as 'the robin imposter'.

Thanks to Ann and Michael Darling for their robin gifts and continued interest in the book, with marked thanks to Michael for the exceptionally generous supply of books, newspaper cuttings and anecdotes.

I am indebted to my parents, Ann and Jim, who supported my interest in natural history from a young age and to my mom who instilled in me a life-long love of learning.

My final thanks are reserved for my partner, Jonny, whose unwavering support, patience and good humour made this book possible and everything else in my life so much better. His willingness to read multiple draft versions at short notice has undoubtedly improved the book.

Robin is dedicated to Esther, who passed away while I was writing the book.

Photo Acknowledgements

The author and publishers wish to thank the organizations and individuals listed below for authorizing reproduction of their work.

Alamy: pp. 19 (Keith Corrigan), 36 (Sonia Halliday Photo Library), 44 (agefotostock), 58 (Nature Picture Library), 88 (mylife photos), 91 (United Archives GmbH), 95 (FLPA), 97 (Cultura RM), 103 top right (Heritage Image Partnership Ltd), 112 (Graham Jones), 113 (Tim Gainey), 116 (blickwinkel), 126 top left (David Chapman), 127 (Paul Brown), 134 (M&N), 149 (David Tipling Photo Library), 150 (Buiten-Beeld), 158 (Minden Pictures); author's collection: p. 103 top left; Bridgeman Images: p. 79 (Leeds Museums and Galleries UK); © The Trustees of the British Museum: p. 43; The British Library, London: p. 22; Brooklyn Museum, New York: p. 40 (Gift of the Estate of Emily Winthrop Miles/ Photo: Brooklyn Museum, 64.98.3_PS1.jpg); buttonmuseum.org: p. 41; Compton Verney Art Gallery/Art UK: p. 55; Düsseldorfer Auktionshaus, Germany: p. 151; Elena Garcia: p. 18; Getty Images: p. 108 (DEA/ICAS94); Houghton Library, Harvard University: p. 16 (AC85 B9345 911S); INT Photometric Hα Survey of the Northern Galactic Plane – Nick Wright, University College London: p. 118. Image based on data obtained as part of the INT Photometric H-Alpha Survey of the Northern Galactic Plane, prepared by Nick Wright, University College London, on behalf of the IPHAS Collaboration; copyright by Philatelie Liechtenstein: p. 100 top left; Los Angeles County Museum of Art: p. 46; Metropolitan Museum of Art, New York: pp. 45, 53, 101; Museum Boijmans Van Beuningen, Rotterdam: p. 86; Natural History Society of Northumbria (NHSN) Archive Collection: p. 20; Nature Picture Library: p. 114 (Ross

Hoddinott); POST Luxembourg: p. 9 (Authorisation number OT/131q21); Walter Potter: p. 67; Rijksmuseum, Amsterdam: pp. 29, 70 bottom, 74; Royal Collection Trust/© Her Majesty Queen Elizabeth II 2022: p. 123; Royal Mail: pp. 129, 130 top left and top centre left; © Royal Mail Group Ltd, 1946 (GPO): p. 132 (courtesy of The Postal Museum); Royal Mint, London: p. 104 top left; Shutterstock: p. 100 bottom left (nadi555); Smithsonian American Art Museum: p. 90 (Gift of Eugene I. Schuster, 1995.82.2); Alex Spade: p. 70 top left; © Victoria and Albert Museum, London: pp. 52, 68, 131; www.visitstockholm.com: p. 147; Wellcome Trust, London: p. 136; Worcestershire Archive and Archaeology Service: p. 135; Yale Center for British Art, Connecticut: p. 8; Fondazione Federico Zeri: p. 152 (this photographic reproduction was provided by the Photo Library of the Federico Zeri Foundation. The property rights of the author have been met.); Kaleel Zibe: p. 12.

© Francis C. Franklin/CC-BY-SA-3.0: p. 6. Francis C. Lincoln, the copyright holder of the image on p. 6; William H. Majoros, the copyright holder of the images on pp. 10 top, 32 and 82; Philip Heron, the copyright holder of the image on p. 14; Anilkumar, the copyright holder of the image on p. 25; Didier Descouens, the copyright holder of the image on p. 94 bottom; and Leon Berard, the copyright holder of the image on p. 145 centre, have published them online under conditions imposed by Creative Commons Attribution-Share Alike 3.0 Unported Licenses. João Manuel Lemos Lima, the copyright holder of the image on p. 10 centre; Charles J. Sharp, the copyright holder of the image on p. 13 (Source: www.sharpphotography.co.uk.); JJ Harrison (www.jjharrison.com.au), the copyright holder of the images on pp. 33 and 34 bottom; Ruby Feng, the copyright holder of the image on p. 94 top; Soumik378, the copyright holder of the image on p. 140; Deepak Sundar, the copyright holder of the image on p. 142; and El Golli Mohamed, the copyright holder of the image on p. 155, have published them online under conditions imposed by a Creative Commons Attribution 4.0 International License. patrickkavanagh: p 33 top, this image was originally posted to Flickr by patrickkavanagh. It was reviewed on 27 November 2018 by FlickreviewR 2 and was confirmed to be licensed under the terms of the CC-by-2.0 (Creative Commons Attribution 2.0 Generic

License). Tokumi Ohsaka, the copyright holder of the image on p. 47, has published it online under conditions imposed by a Creative Commons CCO 1.0 Universal Public Domain Dedication TuckDB Ephemera have published the image on page 65 online under conditions imposed by a Creative Commons 4.0 International License (CC BY 4.0). Glasgow High Street mural, by Sam Bates, aka Smug: p. 71, this image was originally posted to Flickr by Paul Trafford and was reviewed on 5 January 2020 by FlickreviewR 2; Jacob Spinks from Northamptonshire, England: p. 75, this image was originally posted to Flickr by Wildlife Boy1 (Need to Catch Up) and was reviewed on 3 December 2014 by FlickreviewR; Tim Sträter: p. 76, this image was originally posted to Flickr by Tim Sträter and was reviewed on 13 May 2011 by FlickreviewR; Charlie Jackson: p. 83, this image was originally posted to Flickr by chaz jackson and was reviewed on 1 March 2020 by FlickreviewR 2; and Melissa McMasters: p. 80. This image was originally posted to Flickr by cricketsblog and was reviewed on 17 November 2018 by FlickreviewR 2; and all were confirmed to be licensed under the terms of the cc-by-2.0 (Creative Commons Attribution 2.0 Generic License). Wouter Hagens, the copyright holder of the image on p. 126 bottom, has released this work into the public domain. Historisches Museum der Pfalz, Speyer & HMP Speyer: p. 139; (Licence 3.0 Unported CC BY-NC-SA). (Source: www.instagram.com/p/B2YlvZ4iWH9). Adrian Scottow from London, England: p. 145 top, this file is licensed under the Creative Commons Attribution-Share Alike 2.0 Generic License. This image was uploaded by Snowmanradio. This image, originally posted to Flickr, was reviewed on 4 November 2010 by the administrator or reviewer Flickr Upload Bot (Magnus Manske), who confirmed that it was available under the stated license on that date. Arjan Haverkamp: p. 159, this file is licensed under the Creative Commons Attribution 2.0 Generic License. This image, which was originally posted to Flickr, was uploaded by Flickr upload bot on 30 January 2010, 19:05 by Snowmanradio. On that date, it was confirmed to be licensed under the terms of the license indicated.

Index

Page numbers in *italics* indicate illustrations